Jesus The Mystic

Kyle Hockersmith

Evolve Your Mind Publishing Company

JESUS THE MYSTIC

Important – Please Read First

While the practices, disciplines, and insights shared in this book may be beneficial to many, the author and publisher do not provide specific medical, psychological, emotional, or spiritual advice. This book is not intended to diagnose, prescribe, recommend, or treat any particular medical, psychological, emotional, or spiritual issue. Since every individual has unique needs, the content of this book cannot address specific personal circumstances. It is strongly advised that readers consult a licensed and qualified physician, therapist, or other competent professional before beginning any program related to treatment, prevention, or general health. Furthermore, readers should seek medical advice from a healthcare professional before attempting any of the practices discussed in this book.

The information contained in this book is for educational and informational purposes only and is not intended as a substitute for professional medical, psychological, or therapeutic advice, diagnosis, or treatment. The author and publisher do not provide medical, psychological, emotional, or spiritual advice, nor are they licensed medical or mental health professionals. Readers should always seek the guidance of a qualified healthcare provider or professional with any questions regarding a medical condition or mental health concern. Do not disregard professional advice or delay seeking treatment based on the information provided in this book. The practices and concepts discussed are intended as general guidance and may not be suitable for everyone. Before engaging in any practices or techniques mentioned, consult with a licensed healthcare provider or other qualified professional, especially if you have existing medical or mental health conditions. The author and publisher expressly disclaim responsibility for any adverse effects, outcomes, or consequences resulting directly or indirectly from the use or application of any information contained in this book. Use of the content is at the reader's discretion and risk.

Paperback ISBN: 979-8-9922703-0-3
eBook ISBN: 979-8-9922703-1-0

My heartfelt desire is that you continue to seek
Those ever loftier realms of spirituality,
And that you endeavor to truly know thyself through meditation

CONTENTS

1

INTRODUCTION

I am truly honored and grateful to you, the reader, for taking the time to read this book. It is my intention with this writing to provide as thorough of an approach as possible to the concept of Jesus The Mystic, and will do so through available resources at the time of this writing as well as through my own meditation practice. I feel it is important to note that I will not be referring to Jesus in any form of religious manner, in fact, I feel it is quite pertinent to remain secular. I will from time to time utilize the King James Bible but as a reference material and in an allegorical manner.

Let's dive into a early decision I had to make during this writing. You will notice that the title of this book is "Jesus The Mystic", while I rarely use the name Jesus in the chapters that follow. Through thoughtful meditation and research, I was guided to use Yehoshua as the name in place of Jesus. Let me explain my thoughts on this decision. For years, I have been researching the history of the Bible, and the various translations as well as the persons responsible for such canonization and translations. It has become apparent to me that the Greek translation of Jesus for the name of Yehoshua is not correct. The name Jesus is more tied to Greek mythology than to the mystic man called Yehoshua. In Hebrew, Yehoshua was a common name used, and more specifically in the Galilee region.

I will ask that the reader not become overly concerned of the name I have chosen to refer to the actual man, rather, follow me on a journey of remembrance into the mystical realms of his life. As Yehoshua said in the gospels, John 3:16, "Verily, verily, I say unto you, He that believeth on me, the works that I do shall he do also; and greater works than these shall he do; because I go unto my Father." This begins the over arching concept of this book. In concise wording, we are able to do exactly what Yehoshua did, and more. Yehoshua came to show us how to live, and to bring us into remembrance that God or source is within us. When a person learns how to tap into the source within them, the mystical realms begin to open to them and they learn exactly how Yehoshua was able to say the quoted verse above.

At this time in the introduction I feel it is also pertinent to define some terms that I feel are appropriate for this writing. You will find that I use God, Universe, and Source quite interchangeably. As I write, I am in a state of flow, and choose the term as I feel appropriate for the content with which I am writing about. God, Universe, Source, Heavenly Father is describing the ultimate source of knowledge, power, and Divinity. It is what created everything, it is everything, it is the source of wholeness, and oneness. Ultimate love like never felt before exists therein. The Bible, in Genesis, in an allegorical manner quite clearly illustrates the fall from grace, or oneness. From this point forward, we have felt the feeling of separation and have desires to return to oneness or complete wholeness. I can assure you that Yehoshua was bringing those who will hear him into remembrance of this, that we have the ability to return to oneness and complete wholeness.

I have truly enjoyed reading back through the Bible, and other books that I have previously read. I see and understand things so much clearer and new insights are being revealed to me. I have realized that this is truly due to a new level of consciousness. As we practice the lessons that Yehoshua taught, meditate, and strive to move closer to wholeness and oneness, our consciousness is awakened further and our perception is modified such that we see things and understand things more clearly. I love that verse in the Bible referring to *seeing things dim now but later we will see clearly*. To me, this verse clearly exemplifies what I am sharing with you here. Additionally, I have recently been enjoying a new level of understanding with regard to Yehoshua saying in the gospels, *that we must be born again in order to enter into the kingdom of Heaven*. I now realize, that this *born again* process has been happening numerous times, and will continue to happen as I grow and evolve. Each time a person rises up in consciousness, or has new consciousness awakened in them, they truly are seeing the world through new vision or perceptions, a new set of eyes, just as a new born babe would be looking through new eyes when they are born into this world.

I would like to express a certain style of writing that I selected for this book. My desire was to cover the content provided, and new conscious awareness that I have received in a manner that was as easy to read and follow while not getting to concerned with proving through other sources. Rather, I would much encourage the reader to just read with an open mind and use there own intuition to reflect on the words they are reading. I call this checking for heart resonation. Does what you are reading come to you in a sense of *knowing* that this feels correct, even though it may come up against your *programs* or *beliefs* you have been told or taught. Even when it may not resonate, and you wish to refute something within this writing, I loving encourage you to meditate on it, and research for your own interest. Additionally, as I encourage the reader to continue to research topics within regardless if you agree or disagree with them, I as well will continue to research these things. It truly is a daily process of evolution.

Perhaps, lastly, I should share the reason for writing this book. I am a person who is always trying to calm situations, keep things in harmony. So when thinking about writing this book, which has a potential to be quite contentious amongst readers of varying levels of conscious awakening, you can imagine my anxiety. Therefore, I must share the story herein for the curious reader. I have been practicing meditation for several years now and average around one to two hours a day. When I meditate, I am seeking Source, or connecting to the Universe, and in so doing, I have visions and intuitive information given to me. As part of my practice, I attend meditation retreats where we meditate numerous times a day for many days in a row, sometimes totaling 35 hours or more of meditation over several days. It was during one of these meditations at a retreat that I heard a voice entering at an angle to my right hemisphere of my brain. I could tell that this voice was different than what any other person would describe as that voice of reasoning we hear in our heads most of the time. The words I heard were simple and concise, it said, *"Kyle, you are going to write a book"*. You can imagine, I was shocked, surprised by the voice firstly, and then secondly by the specific message. Writing a book was the furthest thing from my mind at that time or really at all in my life. Before I was able to even formulate a question in return to the words spoken to me, the same voice returned and gave me the title for the book. It said, *"And, the title will be Jesus The Mystic"*. There was no refuting the authenticity of this message in my heart, it resonated, and I knew I had to listen.

As the days went by, I was becoming increasingly weary of starting the book. I had feelings of *why me*, and *I am unworthy*, you know those types of thoughts and feelings. About two months elapse from the initial message, and I am letting this message get drowned out by daily life and responsibilities, putting *"it"* on the back burner as they say. One day, while driving my vehicle on the highway, I can still remember exactly the place I was, and see the setting perfectly clear in my mind, the voice returned. Same voice, same location and angle to my head. The message this time was more stern to be honest, and I was left with the reaction of, *"OK, I am on it."* The words that came to me this time were as follows, *"This is not an option, it is an order."* Quite frankly, it left me a little disturbed and ashamed as I had truly been ignoring the earlier message. So I proceeded to begin my work, and I am truly amazed at the knowledge and wisdom that has been revealed to me in so many different ways as I write this book.

With that, I simply wish you, the reader, much Love and Light as you evolve your understanding and allow oneness and wholeness to elevate your being.

2

WHO IS A MYSTIC

I t seems very appropriate to begin this writing with a chapter on the mystic and what this entails. Lets review some definitions first. As an adjective, mystic would be defined as a mystical sense, or as relating to mysteries and esoteric rites. Occult is a word that has negative light attributed but rather one should see the word occult as simply hidden. Additionally, the mystic adjective can be viewed as obscure, or to induce the feeling of awe and wonder. Lastly, the mystic adjective would be comprising of magical properties.

Mystic as a noun would be simply defined as a follower of a mystical way of life and one who advocates the theories of mysticism. This definition of a mystic as a noun moves one to further review the definitions of mystical and mysticism.

Mystical, an adjective, is defined as "having a spiritual meaning or reality that is neither apparent to the senses nor obvious to the intelligence", and, "involving or having the nature of an individual's direct subjective communion with God or ultimate reality."

Mysticism, a noun, is defined as "the experience of mystical union or direct communion with ultimate reality reported by mystics", "the belief that direct knowledge of God, spiritual truth, or ultimate reality can be attained through subjective experience (such as intuition or insight)", and, "vague speculation: a belief without sound basis, a theory postulating the possibility of direct and intuitive acquisition of ineffable knowledge or power." All the above definitions quoted were cited directly from the Merriam-Webster dictionary.

So, who is a mystic then? Well, as we conclude from the above definitions, a Mystic is one who is:

- A person who practices, or lives a life of mysticism and advocates mysticism;

- Who defines mysticism as a belief that we can obtain knowledge and power from sources outside of our awareness;

- Who daily practices a lifestyle of communion with God, and ultimate reality;

- Who believes in learning and advocating, or teaching others, a way of life and knowledge that is hidden, or occult.

This may challenge your present belief. Current culture tends to put a negative light on things that are hidden, or occult. If science is unable to explain, then it is not real. This is changing, and to date of this writing, many of the once occult beliefs are being proven true by the same science that shunned these beliefs previously. Just because something is unexplainable, should not discredit the understanding or belief. Additionally, as mentioned in other chapters, we must develop a sense of "knowing" and this is crucial in verifying intuition and knowledge as we receive it in various forms and formats. I mentioned elsewhere in this writing regarding new vision, or new perspective relating to higher levels of consciousness, and how old information we have received becomes new to us again for we are receiving it with new "eyes".

So now I must bring the title of this book together herein this chapter to illustrate the Mystical Christ, Yehoshua. We will use the King James Bible and the records therein the Gospels to bring clarity to this question: does Yehoshua fit the definitions listed above? Did Yehoshua have daily communion with God, or ultimate reality? Did Yehoshua advocate mysticism, or teach hidden, occult knowledge to others? Did Yehoshua tell others that they too can obtain knowledge and power from sources outside of them and their awareness? The answers to all the above questions is irrevocably, yes. Let's bring in some scripture to illustrate this further.

While never intending to be all inclusive, I wish to highlight several passages of scripture, and then encourage you, the curious reader, to continue the research for yourself into the available passages and records of Yehoshua.

Yehoshua almost always spoke in parables, or stories which are meant to illustrate spiritual lessons. Many times in the Gospels, Yehoshua spoke in parables and even at times the disciples following him were unable to understand. Yehoshua would in that case explain it further to them, saying that the secrets of the Kingdom of Heaven are hidden and only revealed to those who listen to him. I really enjoy this example of one who is in communion with ultimate reality will receive knowledge and wisdom beyond common understanding. Yehoshua was doing this daily, and therefore he was able to advocate, or teach these occult teachings to others. However, he truly valued and respected the process one must go through to evolve spiritually and therefore continued to speak in ways that only true hearted persons with proper intentions would understand. Yehoshua exclaims in the Gospel of Matthew, his great satisfaction that God has hidden this information from the "wise and prudent" and only revealing it to babes. This exclamation made by Yehoshua sends us straight into the passage where he states that we must be born again in order to enter into the Kingdom of Heaven. He is telling us, we cannot enter into this divine wisdom and understanding with our carnal minds, we must elevate our consciousness through daily communion with the mystical, the universe, God, source, and divinity.

Something else I find extremely satisfying for this discussion on Yehoshua being a mystic is the realization that the Torah, the Jewish law of the time, and the Kabbalah, all speak to the reader through hidden truths, only revealed to those who are ready for the information. It is best described as branches of understanding that lead one back to the root. Yehoshua is recorded numerous times referring to trees, roots, vines, husbandry and gardener, to illustrate this concept of branches and roots, which leads us to The Tree of Life, which is recorded in several places of the Bible, however this is also in Kabbalistic text illustrating the flow of energy from infinite source to finite manifestation here in this world. We will discuss the Kabbalistic texts further within that chapter, however, I only wish to point out here that you can begin to see the teachings and wisdom of Yehoshua being connected to these ancient teachings. Yehoshua stated in the Gospels, that he is the vine, and that God is the gardener. I love this illustration of how we, as a person living in this world, are like a plant. A plant has its own consciousness, all we can do to it is try to nurture it with what we feel it needs, air, water, sunlight, nutrients, however, the plant needs to do the work of receiving those said things.

Similarly, we have at our disposal everything that God, Universe, Source, Divinity, has to offer for us to grow and flourish in this life, but we must do the work to receive it. To further Yehoshua's parable, and this is where I feel some get quite stuck, is he is referring to himself as a vine. The reason for this is to illustrate that if we practice his lifestyle and daily actions, we too can be as he was. Why the vine, well, he states elsewhere that we must go through him to enter the Kingdom of Heaven. This is simply saying what I have said above, if we model our life after his, endeavoring to seek the mystical, with daily communion to ultimate reality, God, Universe, Source, Divinity, we too will have the occult knowledge unveiled to us and we will realize the power that is ours to have graciously given by the awaiting Source of all Divinity.

We will conclude this chapter with Yehoshua's statement regarding the law, the Torah. The three sects of Jews at the time of his life were the Pharisee, Sadducees, and Essenes. While it is relatively well known, it is important to note here that the Pharisee and Sadducees were overly obsessed with scribing and the law, they were obsessed with compulsive and almost oppressive observance of the law. The Essenian way of life was deeper, they wanted to live the law, which is why a study into there known way of life will reveal practices and culture that they felt would enhance there ability to commune with Divine. It is herein this understanding that I share the statement of Yehoshua, where he states that he has not come to change the law but rather to fulfill the law. Yehoshua was showing that it is more important to truly live the law then to just blindly obey it. He was also revealing that the law was incomplete, and his life will bring all knowledge and wisdom from Divinity together in one life so that the mystical could be demystified for others to follow after him.

3

TORAH, KABBALAH, AND ZOHAR

T he Torah in simple terms is the Jewish law, and contains the books
of Genesis, Exodus, Leviticus, Numbers, and Deuteronomy. These
books have been canonized into the King James Bible of today, however,
during the time of Yehoshua, would have been kept on scrolls for reading
in the temple.

While debated as to the origins, most believed by scholars is that the
Torah was converted from oral format to writing during the 6th and 7th
centuries BC, during the period of Babylonian captivity of the Jews. It
is believed that the writing of the Torah was so that the Jews would be
able to maintain proper worship while being displaced and with no central
place of worship as their temple had been destroyed by then King Neb-
uchadnezzar. The written Torah would later become a central part of the
religious worship of the Jews with weekly readings from sections of the
Torah, or referred to as *parashah*. This is evident as Yehoshua is recorded

in the Gospel of Luke to have entered into the temple and read from the Torah himself.

The question comes to my mind of, did Yehoshua know of the Torah and had he become knowledgeable of it? I assert a most assuredly yes to that question. Let's discuss the reasons why I feel that way. Firstly, the Torah was the Jewish law at this time, and Yehoshua was a Jew. It is recorded in the Gospels that Jospeh and Mary were of Jewish decent, with the lineage of Jospeh given. It is also recorded in the Gospels that Jewish traditions were upheld by Joseph and Mary, which signifies to me that they cared very much to obey all Jewish laws and customs. In the Gospel of Luke, it is recorded that Yehoshua and his Parents went to Jerusalem for the Festival of the Passover, and it is at this time that Yehoshua, being twelve years of age, was in the temple with the teachers.

Now, who were the teachers, and what were they doing in the temple? These teachers would have been the Scribes who diligently studied the scriptures. The Scribes purpose of study would have been to detect any scribal errors as well as to understand the meaning of the scripture. These scribes would have taken the place of priests.

Yehoshua, finding himself interested in the scriptures, went to the temple in Jerusalem and was recorded to be sitting, listening to the Scribes, and asking questions of them. The record of this experience continues by the statement that the Scribes were amazed by his knowledge of the scriptures, or law (Torah), and his answers. Meaning, Yehoshua was actually teaching the Scribes as well as himself learning from them. This is evidence of Yehoshua having deep knowledge and understanding of the Torah at the young age of twelve. I would continue to express his knowledge of the

Torah by sharing that most likely from birth, Yehoshua was exposed to the Torah through the weekly readings, *parashah*, and customs of the Jews to celebrate the eight Feasts of the Lord. These feasts were to cause remembrance of their ancestors and their experiences.

I would like to divert briefly from the current flow of this chapter to make a few tangential connections. The eight feasts are listed below, however it is important to reflect on the study you may be curious to embrace upon which is the study of how the depicted feasts were to show that Yehoshua was coming in the future. Additionally, I find it curious how we know of eight Chakras in our bodies, and how there are eight feasts. Here are the eight feasts, one of which was a weekly feast, and the dates given are based on a lunar calendar (see chapters on **Kundalini** and **Christ Within** for chakras and lunar calendar respectively):

- The Sabbath Day (weekly feast) – ***ROOT CHAKRA***

- Passover (*Pesach*) - Nisan 14-15 – **SACRAL CHAKRA**

- Unleavened Bread (*Chad Hamotzi*) - Nisan 15-22 – **SOLAR PLEXUS**

- First Fruits (*Yom Habikkurim*) - Nisan 16-17 – **HEART CHAKRA**

- Pentecost (*Shavu'ot*) - Sivan 6-7 – **THROAT CHAKRA**

- Trumpets (*Yom Teru'ah*) - Tishri 1 – **THIRD EYE CHAKRA**

- Atonement (*Yom Kippur*) - Tishri 10 – **CROWN CHAKRA**

- Tabernacles (*Sukkot*) - Tishri 15-22 – **SOUL STAR CHAKRA**

Bringing all of this back together, we can clearly see that Yehoshua was very knowledgable of the Torah, even at a very young age. So knowledgable of the Torah, he was able to teach those very people who were supposed to be the experts on the law, the Scribes. This understanding must first be achieved before you may be able to begin to make the connection to the Kabbalah and determine the profound Kabbalistic connections to the Mystic Christ.

It would be best understood if explained in this manner, the Torah is not separate from the Kabbalah, but rather the Kabbalah is the deepest, most esoteric understanding of the Torah. The Torah is a multi-dimensional text with many levels of understanding, from the written word, all the way down to the most hidden, or esoteric understandings, to which many have spent their lives studying and understanding. The Kabbalah teaches the student that we are created in the "Image of God", and how we are a microcosm of all the God has created. It further teaches the student that we have the ability to create cosmic impacts upon the entire universe through our actions in this life. We will discuss the Kabalah and more specifically the Zohar in deeper manner later herein this chapter, however I would like to move on to a little more history.

The Kabbalah has been referred to as a mystical text primarily because it was reserved for only a select few to obtain and learn of its teachings. Traditionally, as reported from the time of Moses receipt of the Torah, the Kabbalah was transmitted mouth to ear, in an oral format from the master to the student. Only a select few persons in each generation were given this right to learn of this knowledge in such oral manner. In brevity, it is told by certain Kabbalistic masters that this knowledge was at one time freely shared, orally as depicted by scripture in the old testament and Abraham, and thereafter for a period of time in history became coveted and kept hidden, shared only as described earlier. Only later in history has the Kabbalah become publicly available, with the Zohar. What will follow in this chapter is a brief overview of the Zohar and pertinent understandings. I do not intend to make this an all inclusive discussion on the Zohar, but rather to attempt to illustrate that Yehoshua understood this wisdom and truly embodied the Kabbalistic teachings in his daily life. Might I also suggest, that in the same manner as Yehoshua became the Mystic Christ, we too shall endeavor to embody the teachings of the Zohar so evolving into more of our Mystic self.

The Zohar, in literary form, is the quest for spiritual enlightenment through the action of studying the text. It is depicted therein, that the action of studying is the highest form of religious behavior, thereby achieving enlightenment and communion with God. It is this study of the Zohar that entails spiritual contemplation, and the desire to understand the relationships between the divine and human realms. Some major themes found in the Zohar would be as follows: the nature of God and the cosmos; creation of the world; the attributes of God or *sefirot*; nature of evil and sin; the revelation of the Torah; the commandments; holidays, prayer, and rituals of the ancient Temple; and, much more.

As a central focus of the Zohar, the nature of God (known as *Sin Son*, or the *endless one*) is illustrated to the reader of the text as expressions of God's being. Thereby, God relates to the world through these aspects. The ten *Sefirot*, or aspects of God are as follows:

- *Keter* (Crown)

- *Hokhmah* (Wisdom)

- *Binah* (Understanding)

- *Hesed* (Mercy)

- *Din* (Justice)

- *Tiferet* (Beauty)

- *Nezah* (Eternity)

- *Hod* (Glory)

- *Yesod* (Foundation)

- *Shekhinah* (Feminine aspect of God) or *Malkhut* (Royalty)

Let's continue with a quote from Rev Michael Laotian, PhD, "The wisdom of Kabbalah teaches a practical method of attaining the upper world and the source of our existence, by realizing our true purpose in life man attains perfect, tranquility, unbounded enjoyment, and the ability to transcend the limits of time and space, while still living in this world." Kabbalah is best described as the science and wisdom that enables a person to not only feel but to know that there is an upper reality, or Ultimate Reality.

As a core foundational principal of Kabbalah, our perception of reality is fundamental to understand. The Kabbalah teaches that we begin our existence as infinite source, and yet in this life we understand our reality in a much more restricted manner due to the five senses. The complete and total reality is divided into five worlds, *Adam Kadmon* (Primordial Man), *Atziluth* (World of Emanation), *Beri'ah* (World of Creation), *Yetzirah* (World of Formation), and *Asiyah* (World of Action). These worlds can be thought of as worlds (levels) of consciousness, closeness or distance from our connection and awareness of this complete reality (Source). The barrier between our perceived reality and these five levels of consciousness is called *Machsom*. *Machismo*, the Kabbalah teaches, is that internal veil of separation between us and the love of our creator. Crossing this veil occurs upon the onset of the desire to seek something more in this life. When we begin to question, what else is there for us here. Below the veil of *Machsom*, we have no sense or perception of the spiritual realms. Kabbalah teaches us how to retrace our steps that we took in descending down from Ultimate Reality to this world, back to Ultimate Reality.

Our perception of our reality is truly what is trapping us below this veil of *Machsom*. As mentioned in other chapters, we are running on programs in our subconscious that create our perceived reality. The ego plays a major role here as well. Without changes to our ego, we will remain below this veil. However, when in perfect comfort with our present reality, no desire exists to modify the ego and change our perception and subconscious programs. When we lose our comfort, begin to become uncomfortable with our present reality, we lose pleasure and begin to desire something more. It is in this uncomfortableness that we move closer to passing through the veil of *Machsom*. When we begin to develop our ability to sense beyond the five senses, we will begin to climb the ladder to the Ultimate Reality. It is this additional sense, with which we develop, that allows us to get beyond our ego and our subconscious programs.

This concept of becoming comfortable being uncomfortable is the principle driving force that moves one through the various levels of desire, from base or animalistic to spiritual enlightenment. Here are the following categories of desire and the evolution of them (note that one plays into the next, and when one doesn't satisfy we move to the next level):

- Sex, Food, Shelter (Basic Animalistic Needs)

- Wealth

- Power

- Knowledge (Science, Religion, Art, Human Achievement)

- Point in the Heart (Divinely placed, only after the above is not satisfying)

As you move from one level to the next in the evolution of desires, you continue to find that the ones previous as well as the current level they are on does not satisfy, even the final and last level listed above, Knowledge. As this knowledge is based on human achievement knowledge, and not that of Source or Divinity. It might be better stated that the four levels mentioned above are worldly desires. Once we have passed through all of these levels, a fifth level of desire emerges, and it is not of the world. This desire is placed into our hearts and is from the Ultimate Reality, and is called by the Kabbalah "Point in the Heart". Unlike the other four, it continues to grow and can bring us into the upper reality.

The Kabbalah teaches us that the upper reality is the exact opposite of our current state in this reality, which is why we are limited and cannot perceive it. As discussed earlier, our ego and subconscious programs are what limit our perception of this upper reality. It is described that the egoic self is the will to receive, while the opposite is the will to bestow. The will to receive is what is truly limiting our ability to transcend into the upper reality. With the will to bestow, unlimited pleasure and joy exist for us but we are unable to enjoy this reality because we are still bound by the will to receive. It is the *Law of Equivalence of Form*, told to us in the Kabbalah, that allows us a means to pass from the will to receive (egoic state) to the will to bestow state.

The Law of Equivalence of Form, in Kabbalah, is the operation to bring every part of Nature to a state of complete balance. This may be observed through inanimate, vegetative, and animate levels, which are already perfected. Humans are operating at a level above the inanimate, vegetative, and animate levels and we are unaware of the laws and do not know how to observe these laws. The human level, being above the afore mentioned levels, is to be operating on a spiritual level as a creation connected to the Upper Forces. We must learn to keep the *Law of Equivalence of Form* in the spiritual environment.

In conclusion of this concept of *Law of Equivalence of Form*, and the movement from the will to receive into the will to bestow, we must look at a couple more things. Kabbalah teaches the student that there are only two things, the Creator and you. What keeps us from communion with the Creator is the will to receive. The Creator is entirely the will to bestow, and as we descended down from ultimate reality, we moved ever further away from this will to bestow and obtained more of the will to receive.

Therefore, as we move closer to perfect *Law of Equivalence of Form*, we move closer to our Creator, which consequently we obtain ever increasing levels of the will to bestow in place of the will to receive.

The next level of learning from the Kabbalah is the conscious transmutation of our worldly desires into spiritual desires. This process is initiated when the Point of the Heart is realized as mentioned earlier. The Point of the Heart is when the initial spiritual seed is planted inside of our heart, called Israel in the Kabbalah, and begins to grow and the worldly desires in the heart are transmuted (called Nations). It is described that as we fell from Source, further and further separation from our Creator, by design, the collective soul was shattered into 600,000 parts. Each of these parts then further divided into 613 parts (the desires of our heart). It is the 613 desires of our heart that we must work to transmute from world desires into spiritual desires. This process along with others, brings together the collective soul once again. For us to truly transform, the Kabbalah teaches that we must take each event as it occurs and first feel our will to receive in the event, and then contemplate on the Creator and the will to bestow. These events are called are referred to in Kabbalah as *Reshimo*, or data commands.

It is said that our entire life, all incarnations, is contained within the blueprint, or the entire plan of operation for our life, called *Reshimot*. The best way to describe *Reshimot* is like a blueprint, of all the items we must experience to change from will to receive to the will to bestow. The process of climbing the ladder from separation, back to our Creator. It is important to note a distinction here, the Zohar states that it is not ourselves that are performing the changes, rather the surrounding light that is performing the change. We must come with a deep desire for change and then the work is done on us by the surrounding light.

The light must be defined further. It is not the visible light spectrum. The Kabbalah describes a screen, or *Masach*, is the tool for use to perceive the light and allows for us to transform the will to receive into the will to bestow. The screen allows us to decipher a difference between our worldly desires and the corrected desire. Now, what I find really exciting, the Kabbalah describes quantum entanglement by how when a desire is corrected in one person, that same desire becomes corrected in all other souls due to the collective soul. This goes back to what was mentioned earlier about the 600,000 fragments then being subdivided into 613 additional fragments, and how each 613 is a piece of the whole 600,000 fragments. A student of Kabbalah will begin to see the love of the creator when they see this process of one corrected desire in one heart automatically correcting in all other hearts. It is then that one will see the process of enlightenment from a place of ultimate love from our Creator. True enlightenment is when the person can see and understand that no longer is life filled with good and bad events but rather sees and understands the real unifying thought that our Creator is behind the event.

Kabbalists refer to the next level of Kabbalah as the *Speaking Level*. It is in this level that you begin to see the bigger picture of the plan for your life and the plan for creation and how everything works together for the greater good. This is yet another continuation of the transformation from the will to receive to the will to bestow. It is in this level of consciousness, we begin to not see events as good or bad, but rather from the will to bestow, or the eyes of the Creator. What is the quality of bestowal that the Creator intends to provide us with this event? It is the desire to understand this question for each event, that allows the light to provide us the answer. When we do not understand the intention of bestowal from the Creator for a particular event (quality of bestowal, quality of giving), we feel suffering or pain from the event. This quality of bestowal is the general law of the universe. All laws of nature, inner and outer, follow this law of bestowal. We must become like this law of life, quality of bestowal. If we do not, suffering is the byproduct.

These events become necessary in your personal evolution, these good and bad events. It also becomes necessary to have moments in our lives when we do not understand nor can see the good in an event, for by that lack of knowledge we must go back to Source, or God, with the intention of learning. Through this process our own nature as well as the nature of our Creator is revealed to us, the will to receive and the will to bestow respectively.

Love thy neighbor as thy self is something we see recorded in the Gospels of the King James Bible routinely. Love thy neighbor as thy self is also a primary, foundational component, of the Kabbalah. In fact, as we were discussing earlier, you cannot truly understand the will of bestowal as the Creator intended, if you do not truly love your neighbor as your self. To better understand what the Kabbalah is really intending us to feel and to have permeating our total being, is that we truly feel the other person. What we have in our life is theirs, and we are not only ok with giving them what ever it is that we have, we desire to give it. The way we act, think, and behave also carries this concept forward, where we do so in a way that considers the other person with such care and concern as we would have for our own well being. Yehoshua is recorded to have stated in the Gospel of Matthew that the two greatest commandments are as follows: "thou shalt love the Lord thy God with all thy heart, and with all thy soul, and with all thy mind, and with all thy strength: this is the first commandment. And the second is like, namely this, Thou shalt love thy neighbor as thyself. There is none other commandment greater than these."

Yehoshua furthered his teachings on love by stating as recorded in the Gospel of John, "A new commandment I give unto you, That ye love one another; as I have loved you, that ye also love one another." However, it is what he is recorded to have said next is what I find most interesting, "By this shall all men know that ye are my disciples, if ye have love one to another." Essentially, Yehoshua is explaining to his disciples that they will be recognizable, set apart from everyone and everything else, by this one act. People "love" all the time in all kinds of ways. So, how could this be such a trademark, as to set them apart from everyone and everything else at this time (I would postulate that it would also set anyone apart in today's time as well)? It is a level of love that surpasses all human desires, the will to receive as we discussed earlier. The Ultimate Love of the Creator, Universe, Divine, Source. When this love is present in ones life, you truly care about everything and everyone. You reach a balance, or harmony within your self and the world around you in its entirety. You begin to see and understand the great Universal energy and the Divine orchestration of all things and how we play our role(s) within this grand Universe. The *Law of Equivalence of Form*.

As Yehoshua is recorded to have stated to the Pharisees in Jerusalem, he did not come to change the law (Torah, Kabbalah, Zohar) but rather to fulfill the law. Fulfillment in the greatest way, to live it, to embody it, to allow the law to permeate his entire being. In so doing, he was bringing life to the written law, exemplifying the true power of the law, and illustrating to others that any person can do the same. Yehoshua clearly illustrated in his life and the statements we have recorded, that it is the will of the Creator that every person would willingly desire to seek this law for themselves, and begin the journey of understanding, loving, and embodying the law in themselves, thereby evolving their consciousness to ever increasing heights of pure spirituality. As the Kabbalah teaches, we are returning to oneness, wholeness, through the rising of our spiritual consciousness, correcting of the desires in our hearts, returning to our Creator and the collective consciousness.

4

YEHOSHUA'S LIFE

I would like to begin this chapter by telling the story of Yehoshua as we know it to be told in the King James Bible and then expand into the missing years of his life. For those of you who may not have read the Introduction, which I lovingly urge you to do so, I am choosing to use the Hebrew name Yehoshua in place of the Greek name Jesus. You can refer back to the Introduction for more information as to why.

Yehoshua was born around 4 BC in the town of Bethlehem. According to the writer of the Gospel of Luke, during the time of his birth, then Roman emperor, Caesar Augustus, required a tax to be collected which forced Joseph and Mary to travel to Bethlehem. So the biblical story sets forth a very pregnant Mary, Yehoshua's Mother, traveling to Bethlehem and needing a place to stay while having birthing pains. Yehoshua was being born. Joseph, Yehoshua's Father, tried to find a room but due to the census ongoing, there were many travelers and no rooms available. Hence, the birth in the stable and laid in a manger.

It was at this same time, that there were shepherds in the fields watching over their animals, most assuredly sheep. These shepherds saw an angel come to them and gave them a message, that Yehoshua had been born and he was in Bethlehem. These same shepherds went to Bethlehem to see Yehoshua, and shared with his parents what they had seen and heard.

As time passed, the story continues with Yehoshua returning with his parents to Nazareth, a small city is the region of Galilee. This is where Yehoshua was given the name of, Jesus of Nazareth, or the Nazarene.

Now, in the Gospel of Matthew, we are told some more details about the young Yehoshua's life and travels. There were wise men from the east who came seeking a young child calling him the King of the Jews. The wise men stated that they had seen a star in the east and have come to worship him. Herod was King of Jerusalem at this time and was troubled by this statement. He asked these same wise men to help him locate the child, but they were privy to what King Herod was ultimately wanting to do, which was to kill Yehoshua. The wise men met with Yehoshua and his parents in Bethlehem leaving them with the three gifts, gold, frankincense, and myrrh. After the wise men had left, Joseph was warned in a dream that they must flee to Egypt. So, Yehoshua and his parents move on to Egypt and live there until King Herod dies. After King Herod's death, Joseph brings Yehoshua and Mary to the town of Nazareth.

What I really enjoy about this story is that Yehoshua experienced many different places to live as a young child. The King James Bible takes us up to age twelve of Yehoshua's life, and in those twelve years he is recorded as living in Bethlehem, Egypt, and Nazareth. What is really interesting to think upon is the cultures of each of these places.

Bethlehem during Yehoshua's life was a small town about five miles south of Jerusalem. It is estimated of a population around two thousand or so at this time. Bethlehem consisted of mostly one and two bedroom homes and was situated a small distance off of the main north-south reads route called the *Way of the Patriarchs*. What is interesting to note, is that Bethlehem in Hebrew means "House of Bread". The lands around Bethlehem are fertile agricultural lands for harvesting wheat and barley, hence the name Bethlehem. We will talk later in more detail in another chapter, however, I would like to point out the allegorical significance that Yehoshua began his life in the "House of Bread" or as one may find the solar plexus region of our bodies. Refer to chapter on the *Christ Within* where I discuss the *Christos* and how the sacral secretion and sacral seed travel in our bodies. I have no evidence of what life must have been like living in this town, but I get the feeling it was quiet and peaceful living. An occasional passerby traveling along the nearby trade route, however, being so closely positioned to Jerusalem I get the feeling travelers would just pass by Bethlehem.

Egypt would have been quite the change of culture for Joseph and Mary. According to biblical research and review of the scriptures, one can deduce that Yehoshua would have been around two years of age at the time they began living in Egypt. At this time, Egypt was considered a province of Rome, which occurred in 30 BC. Sobeing a province of Rome, Roman Emperors were the Pharaohs of Egypt until later conquests of Egypt, beginning with Augustus. Life in Egypt at the time of Yehoshua living there would have been much more culturally diverse compared to Bethlehem. There were a number of Jews living in Egypt at that time and they were respected by the Romans. It is not well documented but given with cultural norms of the time, Joseph and Mary would have most likely chosen to live amongst a small community of Jews in Egypt.

Hellenism, a practice centered around polytheistic and animistic worship, worshiping of greek gods and spirits of natures, undoubtedly was present in the area that Yehoshua was living during this time in Egypt. I can only further imagine, as parents would do, attempt to shelter the young child from these cultural differences, attempting to raise the child within there own cultural norms. What I find interesting, is that Yehoshua most assuredly would have learned of these practices just by listening to those around him as it is almost impossible to totally shelter one from the surrounding environment and culture.

Lastly, prior to Roman conquest of Egypt, the Egyptian culture was that of love for life and the land. Religion was a part of every day life for every Egyptian. They considered themselves to be co-laborers with the gods, and to celebrate and give thanks for everything the gods had accomplished for them. Egyptian culture is also credited with many firsts and advancements in our day, such as glass working and metallurgy, construction methods, and metaphysical knowledge. In reference to metaphysical knowledge from Egypt, the eye of Horus found in Egyptian archeology, is directly referring to the pineal gland as mentioned in other chapters in this book. Egyptian knowledge of the metaphysical and the mystical was profound in the days leading up to Yehoshua being there. Now, granted, the Roman culture was pervading the Egyptian culture over time since the conquest but it is highly likely Yehoshua was exposed to some of this knowledge in the time that he lived in Egypt.

Yehoshua would leave Egypt and move to Nazareth with his Parents around the age of six to nine years of age. Nazareth was a small town in Israel, or more like a tent encampment at the time of Yehoshua living there. According to some, only several hundred people were living there during this time and were living simplistic lives as farmers and tradesmen. Excavations have proven the size of the town to be very small, efficiently utilized, with cisterns and one spring for water. The culture here during Yehoshua's time would have been traditional Jewish culture. As mentioned in other chapters, one of the three sects of Jews during this time was Essenian, which is synonymous with Nazarene.

What I think is really interesting to note here regarding Nazareth is the close proximity to the much larger city of Sepphoris. At this time, Sepphoris would be around 30,000 in population and was heralded as a Mecca for art and culture. Many international travelers would pass through this larger city, and being only about four miles away from Nazareth it is assured that Yehoshua would have spent time therein the city of Sepphoris and obtaining knowledge of foreign cultures, languages and political structures. Additionally, during this time many trading routes were in use, which the *Silk* and *Incense* routes were connected by the *Via Maris* route which went through the town of Nazareth. Yehoshua would have had numerous opportunities to meet and learn from travelers from the far east.

This is where the wise men come back into the discussion. If you will recall, the biblical account refers to them as wise men from the east. During this time the east would be referred to as Asia and Africa. Now while the Bible doesn't give us much information, we can begin to put some of the information together to reveal more on who these wise men were. In the King James Bible the English translation is wise men, however, in the greek translation they are called Magi.

Who are these Magi? Most likely, they were astrologers in the Zoroastrian temples of Persia. Therein, searching for prophecies and omens in the heavens. We can safely assume such as it is recorded that they were observing a new star in the sky, and that they knew where to go, and what it foretold. A further study on just who these Magi were would provide you, the reader, with a much deeper sense of acceptance to my feelings that these wise men were Mystics of their day and time.

So after Yehoshua moves with his family to Nazareth, the King James Bible does not give us much more information about his youth until he returns at around 30 years of age. Some people refer to this as the lost years of Christ. Through my various readings and meditations, I have a "knowing" that he spent time learning from various travelers coming from other countries therein Nazareth. This instilled in him that there was more to life than what he was being told. Call it a passion inside of him to learn and evolve more was growing. At a certain age, age unknown, he would learn under a mentorship with the Essenian Brotherhood, learning the ways and knowledge of the Essenes. He would then later travel the *Silk* trade route to visit India and Nepal. Some reports of ancient scrolls contained in sealed and protected temples throughout this region, state that a man coming from Israel had visited and stayed with them for a time to learn of there ways and there teachings. He would spend most of these missing years abroad, and upon returning begin the final initiation into the Essenian Brotherhood.

What is most interesting around Yehoshua's reported travels abroad is that while the Essenian origin is uncertain, it is clear to see their teachings in many other teachings. The *Zend Avesta* of Zoroaster contains Essenian teachings. The Essenian Teachings however contain fundamental concepts found in *Brahmanism*, the *Veda of Upanishads*; and the *Yoga* systems of India. You can see correlations in Buddha's sacred Body tree and the Essenian Tree of Life. Lastly, the *Tibetan Wheel of Life* reflects the Essenian teachings. Seeing all of the correlations begins to suggest that Yehoshua wanted to learn from the source of all of these teachings rather than receive from just the Essenes.

The relevance of all of this is important and imperative to your understanding. It is my hope that I have effectively illustrated that the life of Yehoshua was diverse with contact to many different cultures and countries. As we develop an understanding in later chapters on the *Christ Within, Kundalini* energy, *Pineal Gland*, and *Miracles of Christ*, it is imperative that we have developed a thorough understanding to the daily life of Yehoshua along with the knowledge and wisdom he gained from his travel experiences.

5

KUNDALINI

Kundalini is a word that has its origins from the *Upanishads*, which are ancient texts from which the foundation of *Hinduism* was formed. It is said that the Kundalini, is a powerful energy within our body. Often times, the phrase Kundalini awakening is spoken of and this phrase merely refers to the elevation of movement of this energy. For many years of a persons life, this energy remains in the lower body, or lowest chakras. It is described in texts as a serpent, or coiled snake. This is interesting to note due to the coiled structure of our DNA, the caduceus for medical symbology and also from the standpoint that a snake sheds its skin and becomes new repeatedly throughout its life. Let's break this down a bit.

Why is the energy for most of a persons life, stored in the lowest points of our body, or first chakras and what are chakras? Chakras are found to be described in some of the oldest texts known called the *Vedas* from India. Chakras also have there roots in *Hinduism* as well as *Buddhism*. *Hinduism* refers to six or seven while *Buddhism* maintains five chakras. We will discuss more detail on chakras later and I will postulate the possibility of more than eight, perhaps nine. Chakras are merely energy centers as simply put. These locations are areas inside your body where energy is concentrated. Some might be more familiar with meridians within the body. Similar concept. I am being moved to put here the following, our bodies have redundancy just as the universe can be found in us and we are the universe. As above, so below. You can go as microscopic as possible and you still find the universe. You go as macro as possible and you still see the micro represented. Self-reflective universe is the only way I can explain this concept. Everything is mirrored. A part is the whole, and the whole is the part. This is why you can find all major meridians in your ear, and also on your foot.

A simple overview on chakras, we will maintain seven chakras within the body and an eighth above your head for the purpose of this discussion. Moving from lowest chakra to highest chakra:

- 1st Root Chakra - at your perineum

- 2nd Sacral Chakra - about three inches below you navel

- 3rd Solar Plexus - at or slightly above your navel

- 4th Heart Chakra - around you heart center

- 5th Throat Chakra - primarily your throat region

- 6th Third Eye Chakra - your pineal gland region

- 7th Crown Chakra - top of you head

- 8th Soul Star Chakra - about eighteen inches above your head within your aura

Lets take a moment, and discuss electrical engineering and current biology in simple terms. Science is currently proving our bodies are more like electronics than ever thought before. Our brains are now being compared to mainframes, or computer hardware, and our subconscious minds as the software on the mainframe. Our bodies are filled with energy that is flowing over our nervous system, pumping our hearts, and performing many biological functions every second of every day we are alive.

I am going to postulate that our bodies were the first resistors and transformers. What is a transformer? It is a device that takes a voltage of one level and changes it to another level of voltage. What is a resistor? It is a device that regulates or limits the flow of electrical current. Here is the biological connection, our spine is the transformer, and the chakras are the resistors.

Now lets get back to kundalini energy. We said earlier that this energy remains in the lowest part of the body for many years of a persons life. We also said that it is a referred to as a snake, coiled. We also said it was the medical symbol of the caduceus. Lets brings all of this together. Each chakra is a different frequency of energy. As you begin in the lowest chakra, you are at the lowest frequency. In order for energy to move to the next chakra the energy must be transformed, changed into a higher frequency. Why then are the chakras resistors? Well they are the regulators of this energy. The energy is not allowed to pass until it is transformed. Briefly here, more details in subsequent chapters, every month when the moon enters your birth solar sign, a sacral secretion occurs in the Sacral Chakra. This is the renewing of the kundalini just as the snake sheds its skin and becomes anew. And lastly, again briefly here, more detail in subsequent chapters, the caduceus. Why is this found on most if not all medical symbology. Some say it represents the DNA. While the DNA is a double helix, a better explanation is as follows. As we live our life, we experience things and have traumas. If not properly dealt with in the moment of experience, the ultimate outcome is energy becomes stuck or stored in one of the chakras. This causes dis-ease of the body which then later becomes disease if the dis-ease is not abated by release of the stored energy within the chakra. As the serpent rises, or the kundalini energy is elevated, transformed and allowed to enter the next chakra higher, new revelations will be made known the the person. The person will have memories and revelations of feelings and beliefs and other malaise that is stuck/stored in that respective chakra. Those must be dealt with in a patient, systematic process until all is cleared from that respective chakra and then the energy is transformed yet again moving up the next higher chakra. This process continues until true enlightenment is achieved. This is a long way of explaining that by simply allowing the process of raising the serpent, kundalini rising, to evolve within you, you are healing your body. Now does the caduceus make more sense?

Do not rush this process, raising of the kundalini. Let nature do its work. Can you rush a plant to grow? No. Doing so will undoubtedly harm or kill the plant. I was just thinking about how a palm tree grows. What a great example of raising the kundalini within you. Did you know that a palm tree gets its height and its trunk through a birth, death and rebirth process? Each year depending on the species, maybe several times a year, new palm fronds grow out the top and the previous fronds fall down and eventually die. The death of those same fronds that bring life create overall growth in the palm tree. As we work through each chakra, and the energies and traumas stored there, there is this same birth and dying process that occurs. Just as it is impossible to go from a one foot palm to a ten foot palm, it is equally impossible to raise the kundalini from the root to the crown overnight. The healing process must occur in each chakra for true kundalini rising and enlightenment.

Did Yehoshua (Jesus) know of this kundalini energy? What immediately comes to my mind is the scripture talking about the corn of wheat. In the gospel of John 12:24-26 Yehoshua is telling a parable where he tells us that we must daily die and we must hate our life in this world. So doing we will have life eternal and will be most fruitful. Just as was stated earlier regarding raising the Kundalini, daily working through our old selves, allowing our old selves to die (working through and letting go of our traumas and dis-ease), so we may be reborn is the rising of the kundalini at its core. Why does he say we must hate our life in this world, well, if you loved your life you would have no encouragement to work through all your past traumas and allow them to be healed and ultimately die so that a rebirth could occur.

How would Yehoshua have known about the kundalini? Enter the Nazarene, Essene, and the Jains of India discussion. In the King James Bible we read about Yehoshua being a Nazarene. While it is still argued and not provable as to if a City Nazareth existing at the time of Christ, that is irrelevant for this question on kundalini. What is not arguable is that being a Nazarene was a way of life, a practice. Who were the Nazarenes? They were comprised of Essenes. During the days of Yehoshua, there were three main Jewish sects active, the Pharisees, Sadducees, and the Essenes. It is best thought of Nazareth as a small encampment of Essenes during the time of Yehoshua. Nonetheless, the Essenes trace their lineage back to Shem, Noah's son and the dead see scrolls (written by the Essenes) give credit to the Wise Men of India for their knowledge on medicine. Additionally, the way of life and practices of the Essenes, Nazarenes, and Jainism are almost identical. In effort to not recreate many other research documents available, doing your own research will allow you to find many different ways of connecting the Essene/Jainism. This connection back to India is important as kundalini has its origins from the *Upanishads*, which are ancient texts from which the foundation of *Hinduism* was formed.

I am compelled to go back to the serpent. Some may have trouble with this as the King James Bible refers to the serpent as the devil or Satan. More specifically the story of Adam and Eve being fooled by the serpent to eat of the fruit of the tree of good and evil. After so doing, Adam and Eve realized they were naked and ultimately they were banished from the Garden of Eden. Let's, for a moment, look at this story from a different perspective. If the serpent is the kundalini, and once it begins to rise, we become more aware of our past (emotions, traumas, dis-ease) which is very uncomfortable to go through. Most people would agree that it was better to push those feelings down than to have felt them come up and dealing with them appropriately so as to fully release those feelings, thereby letting them go. Kind of sounds like being banished from the Garden of Eden to me. Leaving the nice, cozy, comfortable place for a less comfortable place. Who was responsible for this occurring? The serpent. What happened, Adam and Eve became aware, or awakened. As the kundalini begins to rise, we naturally become more aware of ourselves.

You may ask the natural question then, when do we get back to the Garden of Eden? Will it always be like this? Lets look back at Yehoshua. He said follow me, right? What he was really saying was live your life like I am. In answer to the afore mentioned question, his life was one of learning, evolving, service, but what stands out is his time in the wilderness. While the King James Bible states forty days and forty nights as the duration of this experience for him, I like to think of it as the death and rebirth process that was not just limited to this definitive period. He was fighting with his old self and getting victory over the old self so he could be reborn. Ultimately, he returned from this wilderness experience more powerful than before. Each time our old self is made "aware" to us, we allow it to die so we can be reborn and through this process we are walking back to the Garden of Eden. We reach the Garden of Eden when we have reached true spiritual ascension. The promised land.

Another example I am thinking of is Moses in the King James Bible. This is yet another story where you can see the kundalini energy rising and the ultimate outcome. In brevity, Moses was a young baby when he was separated from his mother. Sent down river due to the mother fearing her child would be killed. Found by an Egyptian woman of the Pharaoh's house, Moses was taken in and raised as her own child. At forty years old, Moses left Egypt and spent forty years in the wilderness. The last forty years he spent of his life, he was consumed by freeing his people from bondage to Pharaoh and traveling to the "promise land" through the wilderness. Now, I am not sure if you caught that, at forty years old he left Egypt (or the new age world of that time). Yehoshua says we must lose our life to gain eternal life. Moses was at the typical midlife crisis age when he left everything he knew was comfortable. He was in the Pharaoh's house nonetheless. But he left and went to the complete opposite conditions, desolate wilderness. This story just resonates with me, a depiction of the story of each of our lives as we honor, accept and work through the kundalini rising in each of us.

The question of why would you want to do this comes to mind. My only answer is very succinct, there comes a time in everyones life when we ask the ultimate question, is there something better than this? Once we ask that question, we realize that we have been living a life of being asleep and it is upon such realization that the awakening process begins. While hard to go through, it is the true path of enlightenment.

6

THE VIRGIN BIRTH

As I was putting this book together, I received an intuitive thought that I should include a chapter on the "virgin birth". We are discussing the Mystic Yehoshua herein this book, and undoubtedly, you may ask the question about what we read and understand of his birth in the Gospels of the King James Bible. Additionally, I have mentioned in other chapters, the recorded statement of Yehoshua, that *we too can do everything he has done and more so*. Let's endeavor to try to understand this concept of "virgin birth" and how it may pertain to us as well.

To begin this investigation in the "virgin birth" of the Mystic Yehoshua, we must first decide if he was a man at all. Then, we must further decide if he was born (conceived) of a virgin mother in the literal sense, or if he was born (conceived) as a regular child is. The Gospels of the King James Bible tells us several different accounts of Yehoshua and his divine conception. What is interesting to note, is therein the Gospels, there are accounts of the lineage of Yehoshua only from the Father's lineage. From a purely literal reading of these accounts, the authors primary purpose is to establish that Jospeh, Yehoshua's Father, is in the direct lineage of King David, and all the way back to Abraham. I would postulate the question, if Yehoshua was divinely conceived of a literal virgin mother, then he would contain no hereditary rights to the lineage of King David, since his father, Jospeh, did not contribute any DNA in the conception.

The Gospels of Matthew and Luke tell of the birth of Yehoshua, where as there is no mention of the birth in the other two Gospels. Scholars agree that the original writing of the Gospel of John predates the Gospels of Matthew and Luke by almost two decades, and interestingly to note, the Gospel of John does not write of the birth of Yehoshua. We do read of the statement made by Yehoshua in the gospels though that we must be *born again* in order to enter the Kingdom of Heaven. Also, he states that we must be like a *little child* in order to enter into the Kingdom of Heaven.

For the purposes of this discussion, the Mystic Yehoshua, let's postulate based on the evidence supporting above, the statement that Yehoshua was a physical man, born in a literal sense of DNA from both parents, conceived in the normal manner of conception between a man and a woman. Later, at some point in his life, we was *born again*, this "second birth" would be referred to as the *virgin birth*.

Before we move into this concept of *virgin birth* utilizing ancient India practices, I would like to discuss the two afore mentioned statements Yehoshua. Earlier I wrote the two statements of being *born again*, and being like a *little child*, in order to enter the Kingdom of Heaven. Both of these statements while different in the first part of the statement, both are the same result or desire, to enter the Kingdom of Heaven. So we must ask the question, metaphysically, what is the Kingdom of Heaven? In all pretenses, the Kingdom of Heaven is above us, of highest spiritual order or convention. What is the purpose of our life according to the Kabbalah, to rise back to Oneness or Ultimate Consciousness from which we fell upon being created by the Creator. The Kingdom of Heaven can be easily thought of as the realms of spirituality beyond the veil of our personal reality, our five senses. We discuss in this chapter as well as others about transcendence, and elevating our consciousness to cross over this veil and enter into the loftier realms of spirituality, or differently said, entering the Kingdom of Heaven.

Being *born again*, of the *virgin birth*, or to say a birth that is not of this world, implies something different occurs inside of us that allows us to rise into these loftier realms of spirituality. The Kabbalah teaches that when our worldly desires no longer satisfy us, we begin to desire something more and thereby reach the *Point in the Heart* phase of our spiritual evolution. It is at this Point in the Heart phase that a spiritual seed is planted in our heart and the process of spiritual evolution begins.

Being like a *little child* in order to enter the ever loftier realms of spirituality (entering into the Kingdom of Heaven) refers us back to meditation. In other chapters, I discuss the process of our subconscious brain, and I discuss how this may be altered through meditation. What I would like to point out here, is that scientifically proven, a child is in theta brain wave states until an average age of six to seven years of age. Theta brain wave states, as we discuss in the Chapter on Meditation, is the gateway into the subconscious because it is a state of hypnosis. It is in this brain wave state that it becomes physically possible to exit from our carnal mind (left brain) and enter into the seat of Ultimate Consciousness (our right brain). I will say the last statement in a different manner for expression sake; meditation allows a person to change the brain wave state into a state that allows the person to enter into the Kingdom of Heaven (loftier realms of spirituality). The same brain wave states a child is in for the first years of their life. This understanding brings a whole new perspective to the Mystic Yehoshua teaching in the Gospels that we must be like a *little child* in order to enter into the Kingdom of Heaven. Yehoshua is instructing us to meditate.

In India, there is described the system of Kundalini, and the concept of the psychological centers up a persons spine. These first three psychological centers represent the psychological planes of concern, consciousness and action. The first being the root chakra, represented by the coiled serpent, or in anatomy the esophagus. Through which we feed our bodies nourishment, which feeds our mortal bodies. The second center, would be our sexual organs region, or the sacral chakra, responsible for our urge to procreate. The third psychological center would be located at the navel region, or the solar plexus. This would be the consuming of our food, and desire to consume. However, it is not the elementary eating but rather the processing, and ultimate obliteration of the food from its original state into something useable for our bodies. From this psychological center our aggressiveness would derive from, the desire to be forceful and angry to others. These first three centers are described by ancient India as the animalistic centers, or animal instincts, all of which are contained in the pelvic region of our bodies. The next psychological center, fourth center, is the heart chakra, and it is herein that one moves out of the lower three centers or animalistic centers and into the human and spiritual centers.

Upon review of the ancient Indian psychological centers, one will find that there is symbolic forms for each center. For the first center, the symbolic form is the form of the lingam and yoni, or the male and female sex organs in conjunction. This symbolic form is repeated at the heart center, a male and female sex organ (lingam and yoni) in conjunction but this time they are represented as golden. This is representing a second birth, different from the first birth, the primal birth. It is this second birth, symbolized by the golden lingam and yoni in conjunction, the reveals to us the second birth is the *virgin birth*. The birth of the spiritual person, from out of the animalistic person they originated. This second birth, according to ancient India texts, occurs at the heart center, when the person becomes awakened at the level of the heart to compassion and suffering with the animalistic person.

It can be said, that it is at this *virgin birth*, the point at which the spiritual person is born out of the origination animalistic person, that a deity (or God) is born. It was at this *virgin birth* that Yehoshua, the Mystic, was born. It is at this level, level of the heart, once awakened as stated above, that the God in you is born. You become as *Sons of God* from the once held status of *sons of man*. This is why Yehoshua clearly stated in the Gospels, that we can be like him and do the same things, and even greater. This is why Yehoshua also stated that we must be *born again* in order to enter the Kingdom of Heaven. I will postulate that this transmutation of the animal self into the divine self is a continual process of evolution, through the daily practice of allowing the heart level to be awakened. As the heart level awakens, more of our animal self is revealed, and we through compassion and suffering transmute the revealed animalistic nature into divine nature.

It is necessary to refer to this as the *virgin birth* because it is conceived of by the spirit inside of us. It is not of the world, but rather of the spiritual realms. When we reference the Buddha, we learn that it is written that the Buddha was born out of the side his mother at the level of the heart chakra. This being a symbolic birth of the Buddha, a representation of this same *virgin birth*, a spiritual birth. He was born of a natural body as you and I were born. What these accounts are trying to reveal to the readers its that allegorically, each and every person has this self-same animalistic nature in them and has the opportunity to rise above it, at the heart center, and become reborn into the spiritual version of themselves. I would postulate, that while it is an option and free will remains, that this is the one and only purpose of our creation, is to become reborn into our spiritual selves.

In the next chapter, *The Christ Within*, I will explain how this *virgin birth* takes place inside of you based upon a lunar cycle, and how the Mystic Yehoshua exemplified this in his life and teachings. I also discuss the psychological centers, and Kundalini in greater depth in the chapter, *Kundalini*, which is the process of raising our energy out of the lower three animal instinct centers. It was the Mystic Yehoshua who taught a parable to his followers regarding the ten virgins, where there were five wise and five foolish virgins who either had enough oil or not enough oil for their lamps. It was the foolish who did not have enough oil, and the five wise virgins who had enough oil. As we bring all of this information together now, do you find yourself interested to see the words "oil", "lamp", and "virgin" in this story?

As you will read in the following chapters, we all have the opportunity to raise the *Christ* within us, or said differently, raise the *Christos* (oil of anointing) that is divinely placed within us each month. This is again, is a reference to the *virgin birth* spoken of early in this chapter. The five wise and five foolish persons were all called "virgins". This is symbolic of the *virgin birth* that is taking place inside every person, whether they are aware of it or not. Sadly, without preservation of the sacral secretion, the *Christos* (oil of anointing), the *virgin birth* is not realized, and the sacral secretion is not allowed to be converted into oil within us (more specifically in our brains at the pineal gland). This location within us, the pineal gland (the receiver of divine light, or otherwise known as knowledge) is the reason for Yehoshua referring to "oil" in the "lamps". The foolish "virgins" did not preserve the sacral seed, and therefore no oil was produced in their body for the "lamps".

I am being guided to conclude this chapter by discussing the time Yehoshua spent in the wilderness. I mention this account from the Gospels in the King James Bible elsewhere in other chapters, but feel it appropriate to mention here with regard to the *virgin birth* and awakening the spiritual person within us. In the Gospels, we can read that the Mystic Yehoshua spent forty days and forty nights in the wilderness. This wilderness would have been an area in the Judaean Dessert, and this area would have been very desolate. We are also told in the accounts of this time in the desert, Yehoshua fasted the full forty days in the wilderness. It was at the end of these forty days of fasting, we see the account of the temptations of Yehoshua. As one reads these accounts of the temptationof Yehoshua, it is evident that Jacob also went through a very similar process at Pineal. Jacob wrestled all night it states, and then the angel blessed him the next day. Thusly, rewarding him for his struggles and victory. Yehoshua also, was ministered to by angels when he succeeded through his temptations.

Now, what were they victorious over, and what where they struggling with. They were struggling with, and obtained the victory over themselves. Themselves being the lower three psychological centers of the animalistic man. In Kabbalah, the were correcting their wrong desires and learning the will to bestow over the will to receive. The were allowing the *virgin birth* to be preserved and to grow. Why do you think Jacob chose to call the place where he wrestled all night, Pineal? This is where the victory occurs at, our pineal gland. This is the place which the sacred seed travels to from our lower animalistic centers all the way up through the heart and ultimately becomes transmuted in the pineal gland.

The three recorded temptations of Yehoshua are interesting to review in the metaphysical realms, as we see them from the position of rising through the three lower centers. The first temptation, we read Yehoshua was tempted to change the stones into bread, seeing that he was hungry from fasting for forty days. This is symbolic with the first psychological center, the root chakra. This is our root, or base, our primal needs of existence. The second temptation, Yehoshua is tempted to throw himself over a cliff edge to prove that the Universe will protect him. Overcoming our fears and doubts, trusting in the divine is paramount in our belief that the Creator, Source, Divine, Ultimate Universe is always working for our best intentions. This is all symbolic of the second psychological center, the sacral chakra. The third temptation, Yehoshua is shown all the kingdoms of the world and the glory of all of them, and told he could have all of this if he would just bow down and worship the devil (the devil, being that animalistic desire that exists inside each and every one of us). It is this desire which must be overcome in order to rise up the spiritual pathway to enlightenment. Now evidenced herein this third temptation, is the symbolic third psychological center, the solar plexus. Herein this center would be our animal desires to fight, and concur so that we may be victorious and wealthy in the earthly realm with worldly possessions. Yehoshua was able to overcome this temptation by transmuting the desire to concur and fight, with a love for everything divine and spiritual. His love of what is beyond this realm, in the higher loftier realms of spirit.

This is the example of the *virgin birth*. The rising up of our selves from the lower three psychological centers, or chakras, and awakening at the heart level, compassion and suffering for the animalistic person. At such a moment of awakening, through the suffering and compassion, the person may truly be born again into the new spiritual person.

7

THE CHRIST WITHIN

I t was a few years ago when I first heard of this concept. The concept of the Christ is within you. However most people fail to really grasp the true concept of this statement. What does this truth mean? How would or could this be possible? Is it the holy spirit? I find it best to explain it in this manner. Yehoshua states in the gospels, and I paraphrase, everything he can do, we also are able to do and more. So if that is truth, that we are able to perform the same miracles, and heal, and walk on water, and so forth just as Yehoshua did, what allows it to be so. A physical version of Yehoshua inside of us? I think not. Rather, the answer lies in connecting to the true source of power within us even as Yehoshua did.

What is the Christ? The word is of Greek origin *Christos* which means anointed one. Was Yehoshua not anointed with oil? We too anoint our own head with oil through a process called sacral secretion. Buddhism also refers to the sacral secretion as well, *ambrosia*. Hinduism refers to a divine sector of same concept, *Amrita*. In a most simplistic description, at various times, a process biologically occurs inside our body where secretions flow down our spine and are converted into a sacral secretion at the solar plexus. We will break this down further herein this chapter, but first lets reference several stories in the King James Bible that describe this process in an allegorical nature. The scripture reads, "Entering in to the land of *milk and honey*", this being representative of the pineal and pituitary glands. The scripture that details the experience of Jacob and Jacob's ladder, whereby he seeing angels climbing up and down the ladder, is indicative of our spine with the hippocampus, or our brains the kingdom of heaven. Jacob wrestled in the place he called *Pineal*, which is representative of each of our struggles to overcome the lower chakras and emotions/energies associated to elevate all the way to activate our pineal gland. Moving away from the Bible, lets look at the story of Santa Clause. He comes down a chimney to provide gifts which we realize only after he has magically left back up the chimney. Santa Clause comes from the word claustrum, the oil secreted by the cerebrum. Just like the kid is told, if you are good, you will get gifts. If you preserve the claustrum, through being conscious to preserve its holy nature, it will go down your spine, be converted into a sacral seed and rise back up your spine to give you gifts of enlightenment and transformation.

Does this occur for only certain people? No. No one is the exception to this biological occurrence. The difference is realized when you are working consciously to preserve the secretion and allow the sacral secretion to truly rise being converted into gifts as described above. Lets walk through this biological process.

This process, which we will refer to as sacral secretion from here on, is a monthly occurrence synchronized to the lunar cycle. A lunar calendar is based on 28 day rotation which equates to thirteen lunar months. Interestingly to note, a turtle shell exemplifies this 28-day lunar month and the thirteen months per year. As I have become more consciously aware of the world around me, more and more I observe where nature exemplifies these truths we are learning.

Sacral secretion occurs once per lunar month, when the moon enters your zodiac sign. Zodiac for those unfamiliar is based on the twelve zodiac signs or constellations in the sky. It is described as a belt around the heavens extending nine degrees on either side of the plane of the earth orbit. The moon and planets orbit entirely within the zodiac. Interesting to note, the zodiac signs occupy 1/12 or equal share of a complete circle, thirty degrees.

- Aries: March 21 - April 19

- Taurus: April 20 - May 20

- Gemini: May 21 - June 21

- Cancer: June 22 - July 22

- Leo: July 23 - August 22

- Virgo: August 23 - September 22

- Libra: September 23 - October 23

- Scorpius: October 24 - November 21

- Sagittarius: November 22 - December 21

- Capricornus: December 22 - January 19

- Aquarius: January 20 - February 18

- Pisces: February 19 - March 20

Reviewing the lunar cycle each month, one will notice that the moon will enter each respective zodiac sign for about three days. Make note of the three days, as we will discuss this in more detail later. So when the moon first enters your zodiac, the sacral secretion process begins. Within the three days, the secretion will go down your spine, and rise again to your pineal gland. It is imperative that the seed be protected during this period so that it may be preserved.

How does one preserve or protect the seed, and what things inhibit the sacral secretion process? Well, lets go to the King James Bible for some answers. Yehoshua refers to the the greatest commandment of all is to love God with all your heart, soul, and mind. He also says to love your neighbor as yourself. Interested how Yehoshua brings attention to the truth that Love is the most important thing to have in our life. The opposite is hate, negativity, resentment, fear, frustration. There are energies associated with all things in this life, including our thoughts and feelings or emotions. If we are living our life in Love, we will be vibrating at high energy levels, conversely we will be vibrating very low frequencies if we are living in hate or the other low energy feelings mentioned above. Preserving the sacral secretion begins here, working to vibrate at ever higher levels of consciousness by living in Love. The gospels continue to illustrate further things that are derogatory to the sacral secretion process: pride; envy; wrath; gluttony; lust; sloth; and, greed.

Again, I am taking liberty to paraphrase.

One desiring to practice the sacral secretion process, should at a minimum refrain from living a life of negative attributes as mentioned above during the three day lunar zodiac period. However, I would encourage the interested reader to work towards a daily life where these negative attributes are non-existent all the time. Additionally, the things we consume, be it food, liquids, news, conversations with others, make a difference as well in this sacral secretion process. Tobacco, caffeine, alcohol, non-organic foods, fatty foods, foods with much preservatives, all work against this process. It should be said that the sacral secretion process is a life style change. Lets layout the steps below:

- *Step One:* Becoming aware of this process;

- *Step Two:* Developing the desire to see the effects your life;

- *Step Three:* Making a conscious effort to observe the three day lunar zodiac period by choosing a cleaner lifestyle during that there day period (such as eating cleaner, healthier foods, avoiding negative news and conversations, choosing Love over all other emotions or feelings, avoiding use of tobacco or alcohol, etcetera);

- *Step Four:* Repeat monthly observing what happens differently for you each lunar zodiac cycle (observing the small things and changes in your thinking, or insights you may have about the world around you or your own life);

- *Step Five:* Progress towards a lifestyle change that encourages the sacral secretion process and preservation of the sacral seed outside of just the three day lunar zodiac period.

Lets talk for a brief period here about calendars and the history of recording time. The first calendars created where lunar based. At present, a lunar calendar was located in Aberdeenshire and has been dated to be around 10,000 years old. The first calendar to use a 365-day period based on the solar year is the Egyptian calendar. Consequently, the Egyptians also maintained a lunar calendar, where each month began on the new moon. The Egyptian lunar calendar would include a thirteen month every several years, called *Thoth*, to balance the discrepancy between the lunar and solar calendars. It was the Roman emperor Julius Caesar who created the more recent calendar reform that we may be familiar with, the *Julian* calendar.

It wasn an Alexandrian astronomer, Sosigenes, who brought the Egyptian solar calendar with some modifications to Julius Caesar which included the 365.25 days per year and varying days per month, some 30, 31 and February consisting of 28 days. It was much later that the Gregorian calendar was introduced, by Pope Gregory XIII in 1582. One of the most important things to note in this change to the Gregorian Calendar is the shifting of days. The result was a shift in the calendar ahead by ten days. So, what was October 4, 1582 became October 15, 1582. This was proposed and adopted to reckon a regression in the Julian calendar of about one day per century.

All of this information becomes so relevant to the sacral secretion observer when you begin to realize that there is a universal calendar, and then there are man's calendars. Using todays calendar, you may miss the actual secretion period. In order to be a conscious observer, one should begin with locating the exact time, date, and location of your birth. Utilizing internet based astrological calculators, determine your astrological chart which includes your zodiac sign. Secondly, conscious observation of the lunar cycles and in which zodiac the moon is entering is imperative being aware of possible discrepancies with calendars. One practice some subscribe to is a total of seven days observation period, being three days before the expected lunar zodiac period begin date and the three days during with one additional day thereafter for a total of seven days. This is a practice, therefore each individual is encouraged to experiment and learn through this process.

Going back to the physiological process of the sacral secretion, I would like to compare the actual process to what Yehoshua went through in his life. Yehoshua was born of Mary and raised by his earthly parents who were Joseph and Mary. As the claustrum is produced in the cerebrum, the claustrum first passes through the pineal gland and pituitary gland, otherwise referred to metaphorically as Joseph and Mary. It is when the claustrum passes through these glands that it is differentiated. The pineal gland releases a masculine electrical portion of the claustrum which is known as honey while the pituitary gland releases the feminine magnetic portion of the claustrum which is known as milk. This is where the Biblical scripture referring to the land flowing with milk and honey begins to make some sense to our real life and this sacral secretion process. Science has researched the different fluids from each of these glands and has determined that the colors are white (pituitary) and yellow (pineal). It should be noted here that the Egyptians knew of this life preserving process, sacral secretion, and is depicted in the eye of Horus.

When the two secretions from the pineal and pituitary are produced, they flow down the spine through the semilunar ganglia. The secretion is converted into the sacral seed within the solar plexus, otherwise known as the *House of Bread*. So many times Yehoshua referred to the *Bread of Life*, and the story is he was born in Bethlehem, or the *House of Bread*. The Christ is born in each one of us each month when the secretion becomes the sacral seed in our solar plexus. Then this seed must be preserved and raised up the spine back to the pineal gland. I find it interesting that Yehoshua lived 33 years and we have exactly 33 vertebrae in our spine. Remember, I asked you to make note of the three days. Well, Yehoshua was in the tomb for three days and the sacral secretion process is three days. Where is it reported that Yehoshua was crucified? In all four gospels of the King James Bible, he was crucified on *Golgotha*, which in Latin refers to mean the place of the skull. Same as Yehoshua's life, for us, the Christ within happens as a gift from the Divine or Universe *(claustrum, pineal, pituitary)*, and is planted into our solar plexus *(House of Bread, Bethlehem)*. We work to preserve and raise the sacral seed up our thirty-three vertebrae *(Yehoshua's life of thirty-three years)*. The sacral seed completes the successful path by being transformed into true enlightenment at the pineal gland once again, and this process takes three days *(Golgotha, skull, three days in then tomb, rises again)*.

As each sacral secretion that is preserved and allowed to be transmuted into true enlightenment, the conscious observer will grow in knowledge, true beauty *(both inside and outside)* and will truly transform. The Essenes knew this and practiced daily to ensure their bodies were a vessel that would be truly conducive in preserving this seed. What are some of the benefits to consciously practicing observation and preservation of the sacral secretion:

- Knowledge and enlightenment from divinity or source

- Intuition to complex situations in your life

- Natural healing within your body bringing your physiology into harmony

- Youthful vigor and appearance

- Peaceful disposition and calmness in all things

- Enhanced consciousness and understanding of the world around you

There is a connection between the sacral secretion and the Kundalini energy. As we discussed in the chapter on Kundalini, it is referred to as the serpent or coiled energy rising in a helical manner. This sacral secretion, once converted to sacral seed in the solar plexus, travels back up the spine in the same helical manner.

Did Yehoshua practice the sacral secretion observation? I think it is evident in the gospels to say irrevocably yes. We established that Yehoshua was an Essenian. We know from many historical accounts that the Essenes knew of this physiological process and the amazing mystical benefits of it. It is recorded that their daily life style and practices encouraged preservation of the sacral seed. One right of passage for being initiated into the Essenian Brotherhood was spending forty days in the wilderness with no food or water to preserve the human body. Living purely on the preserving effects of the sacral secretion and divine or universal sustenance. We read in the gospels of Yehoshua's journey in the wilderness for forty days and forty nights. Even scripture of him saying that *man shall not live by bread alone but by the word of God.*

We cannot conclude this chapter without a conversation about meditation. The process of preserving the sacral secretion and raising the *Christos* within must include a meditation practice. Meditation shuts off the thinking brain, or as Buddhist would call the *monkey mind.* This is important because the carnal mind, or left brain *(the analytical mind)* is responsible for only ten percent of our brain capacity. True enlightenment occurs when we successfully cross over into the seat of christ consciousness or the right hemisphere of our brains. Pineal activation occurs when we meditate, and when we meditate we are in darkness *(eyes closed).* This allows for melatonin production to occur. Yehoshua said, your eye will be single in darkness. Sounds like third eye activation when meditating in darkness to me. A daily meditation practice will provide the means for the sacral secretion observer to have a pure mind, a familiar connection to universal source, and a body that is in harmony. I will discuss more on meditation in detail in that chapter.

In conclusion of this chapter, Yehoshua was living a life of learning knowledge from sources available to him, which included the Essenian way of life. This included Egyptian knowledge, Hinduism and Buddhism, as well as from Yogi and Magi. Not only did he learn of it, he practiced it in his daily life, which allowed him to daily evolve and grow in divinity which was provided to him from source, or universal consciousness.

8

MEDITATION

In my opinion, this may be one of the most important chapters of the entire book. Herein this chapter we will discuss meditation in light of history, physiological impacts, metaphysical impacts, and how Yehoshua utilized the practice. So let's begin.

Meditation as defined is to know thyself. How does one unpack this definition. To truly know thyself, means to sit with your self and really know your self from your thoughts, to your feelings, hurts, traumas, and all of the inner workings that make you who you truly are. Most of us are living *programs* of our self. Let me explain. From the moment we are conceived we are subject to information coming to us. Science has proven that every child is in a state of hypnosis for many years of their childhood. Typical ages where the child is in a state of hypnosis ranges from zero to seven years of age. Depending on various factors, a child may remain in hypnosis longer or exit this state of mind sooner. Why do I say a state of hypnosis? Hypnotic state of mind is maintaining a brain wave state in Theta. Children in the age range mentioned above are typically in Theta brain wave states all the time. Below you will find the various states of brain wave activity in general terms (from highest to lowest frequency):

- Gamma: 30Hz-100Hz

- Beta: 16HZ-30Hz

- Alpha: 8Hz-15Hz

- Theta: 4Hz-7Hz

- Delta: 0.1Hz-3Hz

Going back to the statement, "most of us are living *programs* of ourselves", it is derived from this scientific fact that as children we all were in a state of hypnosis and were *programmed*.

Programming is an interesting word. We see it for use in describing television *programming*, theater gives out *programs*, software engineers and electrical engineers *program* technology, our computers run on *programs*, and amongst many other examples we use in our common vernacular today "get with the *program*" when communicating our discontent with someone non-conforming. As biologists, psychologists, and other science disciplines dive deeper into the inner workings of the brain and our psyche, we are learning that our brains run on *programs* as well. Surprisingly the common computer system we are so familiar with today, is so similar to our brains and functionality.

The modern day computer has a mother board and central processing unit, it has RAM or random access memory, a hard drive for storage, and software programs, which result in a display on the monitor for user interface. Our human bodies have a brain (mother board and CPU), the short term memory and long term memory is the RAM and hard drive storage respectively, and the eyes become our display (monitor). This can be further exemplified by the reticular activating system or RAS. The RAS actually performs as a buffer of information into our brain seeing that we take in more information than our brain is able to process. The RAS buffers this flow of information using a bias as defined by the person. The most common example of the bias buffering is when our interest is peaked about something, we see it everywhere when before we became interested we never saw it. For example, a particular type of vehicle may peak our interest. All of a sudden, it seems like everyone has that vehicle, when prior to your interest you never noticed. Now these are easy examples of how this bias buffering works, but lets extend this bias buffering ability of the RAS to our *programming*, or software running in our mind. It allows us to view the world we "see" through a bias based on what we know or understand to be truth. This bias is set primarily with the early years of our life while we were in Theta brain wave states. Acting as sponges, we as children absorbed everything we saw as truth, just a way of life that must be true and correct. Unfortunately, not everything we "absorbed" during this period in life is true and correct universally, rather it was true and correct for the moment in time, shall I say respectively true and correct.

Meditation, to know thyself, the true self. As we practice meditation, we become more intimately aware of our truest self and how the *programs* we are running may not be congruent with our highest and best self. This is why I said earlier in this chapter, we are all unfortunately living *programs* of ourselves. The beauty though, is that through meditation practice one can become aware of and familiar with these *programs* so that changes may be implemented.

Subconscious Mind and Conscious Mind

As I eluded to earlier, our brains are like the mainframe and central processing unit of a modern day computer, and we have software or *programs* running in our minds. Science has proven this through brain wave function analysis as well as other techniques. The currently held scientific understanding is that between eight and ten percent of our daily activities are conscious thought activities while the remainder is subconscious thought activity. It is in our subconscious *programming* where the life programs are stored from those early childhood years. Now, I should not be remiss to mention that subconscious *programming* can also be the result of traumatic experiences later in life as well as repeated occurrences throughout our life. We tend to store these types of experiences in our subconscious as a way of preservation.

When we meditate, we are slowing our brain wave state down into Theta and high Delta. This is the door into the subconscious, where we can begin to become really aware of what *programs* are stored there. Once aware, we can begin to make conscious decisions on what we would like to change.

Someone may ask the question, why do I care about the subconscious mind? Well, believe it or not, it is running the show. Case in point: your driving a car, talking on the phone while chewing gum, thinking about what to say next in your conversation on the phone, and all of a sudden you pull into your driveway having the realization that you never thought about your speed, the cars around you, the stop lights and stop signs, how to turn the steering wheel, how to brake or speed up the car, or even navigation to get to your driveway. Your subconscious was running all of those *programs* for you while your conscious mind was having the conversation. It truly is amazing to become so consciously aware of your subconscious mind. So, now as we unpack the power of the subconscious mind, it is time to share that it is not all good up in there. We have *programs* that are working for our benefit but also working to our detriment.

A really good way to see this is to begin to look at big things in your life: your love life, and romance; your financial status; your career; your relationships with your children; and, so forth. Simply put, any thing you are struggling with means your conscious mind is struggling against a subconscious *program*. Anything your conscious is congruent with your subconscious will naturally synchronize and seem so effortless. The goal herein this chapter is to exemplify this fact, when the subconscious mind is truly in line with your truest highest self, everything that once seemed challenging will become effortless.

So what is the goal of meditation? To unlock your truest potential through the unveiling of the subconscious and thereby gaining access to a wealth of knowledge and information otherwise unavailable to anyone who remains in the conscious states.

Types of Meditation

Next, let's define and describe several types of meditation that you may or may not be familiar with. While not conclusive, we will cover a wide variety of practice herein this chapter.

- Mindfulness Meditation: the process of being fully present with your awareness. Some would refer to this as just being present, in the moment, noticing your surroundings and not reactive to what is going on around you. In this process, you simply observe your thoughts and emotions but let them come and go without passing any judgement upon them. Mindfulness meditation may be practiced with eyes open or closed, seated or standing.

- Transcendental Meditation: seated or laying down, to perform this style of meditation, you would repeat a word, phrase or sound, and repeat it for twenty minutes or so per day. This practice is great for finding peace and calmness in your day.

- Guided Meditation: this form of meditation is a guided meditation by someone other than yourself and works to utilize as much of our senses so as to evoke calmness and relaxation.

- Vipassana Meditation: an ancient Indian form of meditation used to see things as they really are. The goal of this meditation technique is to transform the self through self-observation. By focusing your attention on physical sensations in the body, one can establish deep connections between the mind and body. This promotes a balance of the mind as well as love and compassion.

- Metta Meditation (Loving Kindness): While similar to Mindfulness and Vipassana meditation, Metta meditation is the practice of directing well wishes towards others by reciting specific words and phrases to evoke warm-hearted feelings. Typically practiced while seated, after a few deep breaths, words are recited to direct loving kindness to yourself first, then directing same loving kindness to those of who you wish to do so for.

- Chakra Meditation: Meditation practice specifically targeting one of the chakras in the body. Typically one would practice by focusing on the color of that chakra, or may use lights and crystals while meditating.

- Yoga Meditation: while some may think of yoga as a physical fitness type of practice, it actually dates back to ancient India. The postures and controlled breathing intent is to promote flexibility and calm the mind. The required balance and concentration of the practice encourage those to focus less on distractions and stay more in the moment.

Regardless of the type of meditation you choose, the result of proper meditation would be the same, passing through the veil of the left brain into the right brain and touching source.

As I have mentioned before, effective meditation is the key, and to be effective we must quiet our left brain or the analytical brain. This is because the left brain is where the conscious mind is contained, while the right side of our brain is the seat of God consciousness or Universal Divine Knowledge. Yehoshua even said there I shall be on the right hand of the throne of God. When we successfully pass through this veil of consciousness, and enter into the right hemisphere of our brain, a wealth of knowledge becomes available to us. Now, this is a practice, and I will be the first one to tell you that while I may make it sound easy, this is a challenge for most including myself. In the beginning you will naturally move back and forth between these hemispheres of the brain. You will improve your practice of remaining in the right hemisphere longer during your meditation by not reacting to those thoughts that seemingly come from no where, but rather love them and then softly push them away, returning to your meditation. As with anything, the more we practice the easier and more natural these things will be for us to perform.

Another key for an effective meditation is to learn how to and train your brain to move through the brain wave states and remain at the desired brain wave state. We have already discussed the various brain wave states earlier, but I would like to emphasize that Theta and high Delta are the desired meditation states. The reason for this is we loose our analytical state of mind, thereby passing to the right hemisphere of our brain when we enter these brain wave states. However, go to far into Delta and you may fall asleep. It is important to note that we do this every time we fall asleep at night. We lay down, shut our mind off, and we transition from our waking brain wave state at that time (beta or alpha) and then move through Theta into low Delta. I am reminded of the scripture where the disciples were "praying" with Yehoshua and he sees them sleeping. I have always felt that Yehoshua and his disciples were meditating, not praying, and they slipped into low Delta while in meditation. This happens to all of us at one time or another. The important thing to remember is to love yourself in all of your imperfectness. This will make you perfect.

In conclusion of this chapter on meditation, it should be explained that darkness is imperative as well for effective meditation. Scripture in the King James Bible says that in darkness came great light, as well as, pray in secret in your closet. What is it about darkness? I thought everything was supposed to be about light. Well, science has that answer for us, it is melatonin. Our eyes transmit the level of light information to our pineal gland. When the pineal gland is informed of higher levels of light, it produces serotonin, and conversely, when the pineal gland is informed of lower levels of light, it produces melatonin. Now, melatonin is a protein through which the body is able to biosynthesize into other proteins such as Tryptophan, 5-HTP, Serotonin, N-acetylserotonin, Pinoline, 5-MeO-DMT, dopamine, and the *God* molecule, DMT. If interested, you can research further into this biosynthesis process and how meditation practice promotes it. For the purpose of this chapter, increased melatonin production, from being in the dark, is providing your body with extra melatonin for this biosynthesis process to occur and produce the other feel good molecules. The reason I said, the *God* molecule, DMT, is because it is believed that this molecule is responsible for our ability to cross over the realms of human into spirit realms. It is also believed that DMT is the molecule that allows us to commune with higher vibratory beings, or what I would rather refer to as more spiritually evolved beings whom have ascended beyond this plane of demonstration we exist on.

9

MANIFESTING

There are many spiritual modalities today, of which we can learn and practice, and I have discussed several here in this book. The one thing though that I observe as I learn of and study these various spiritual modalities, is a commonality between them. Every modality refers to a Creator, or Source, Universe, Divine. When one really looks into this, one begins to see that there is this common denominator amongst all modalities. Why would this be so, and why would this be important?

Understanding our place in this reality, that we are a divinely ordained and created creature for a purpose allows one to begin to understand the true nature of the Universe. I postulate that it is only then that one can truly begin to understand manifesting. Let's endeavor to describe this in more detail for a deeper understanding.

What is manifesting really, and what is it about? Manifesting in the most simplistic definition would be to create something out of nothing. Or, should I say, seemingly nothing. In the chapter on Yehoshua's miracles, and other chapters, we discuss the importance of being in communion with the Creator as well as in harmony with the Universe, *Law of Equivalence of Form*, and in a state of resonance. We also discussed in other chapters that we are living in a realm that is a holographic projection, and we have an ability to rise above the projection to modify or change the projection we realize in this realm. The only way to be able to truly understand and conduct changes in such manner would be to begin to understand your place in the grand Universe. Truly loving the Creator and the grand design, and embodying a desire to be in harmony with it.

This is where the practice of meditation finds utmost importance. We discuss in the respective chapter on meditation how one learns to "know thyself". In the chapter on Kabbalah we discuss correcting of our desires and becoming in harmony with the Universe, and with our Creator's will. Meditation is paramount in allowing one to really begin to see the errors of their way, or being, and begin to make the needed changes to become harmonized, synchronized with the Universe and our Creator. It is in this harmonization, or synchronization, that true resonance occurs (or to say our frequency of our spirit matches that of the Universe), and when this occurs, wonderful manifestations become our reality.

What I have observed in my own life, is when we become more harmonized, our desires begin to change. We no longer find ourselves desiring the more worldly desires, rather we find our desires changing to that of the Universe. Is this what Yehoshua was feeling when he is recorded to have stated that he must be doing the will of his heavenly Father? Additionally, as we become more harmonized and in synch with the Universe, we begin to develop ever increasing trust-faith-confidence in the Universe. Yehoshua taught in the Gospels of the King James Bible to not worry about tomorrow, and to not care about food or clothing. He implies through those passages that everything works its self out because the heavenly Father cares for us. This teaching of Yehoshua fully exemplifies the statement to follow: become in synch, harmonized to the Universe flow of energy, loving of the grand design, and one will truly see how all things work together in harmony and one no longer fears the unknown but rather trusts in the Universe to forever provide in a manner resonate with the level of the persons spiritual frequency.

Belief

We need to discuss belief for a little bit here as it is the next most important thing in manifesting. If you attempt to make a change in this realm by passing the veil to make a change in the hologram, but return to this realm not believing in the change manifesting, than it will not happen. You must not only believe the change has occurred, but must also embody the feelings that the change has occurred. Why the feelings? Well, the Kabbalah talks about this concept through explanation of the spiritual realm versus this realm. In this realm, we have time and space, through which we must travel and exert effort to implement a change. In the spiritual realm, the Kabbalah explains that it is all feelings that either repel or attract. We return to the term resonance again. I have stated in other chapters that everything is energy, and frequency. Frequency would be most easiest defined as the rate of pulsation, measured in hertz, and resonance would be most easiest defined as being resonant or in sync with the frequency. The Kabbalah shares with the student that as we want to draw closer to the ever increasing spiritual realms, we must understand that our frequency must be resonant with the spiritual realm frequency. It is our feelings that emit the frequency. Therefore, it is our feelings that must be resonant, or in synch with the feelings of the Creator, Universe, Source, Divine.

The Mystic Yehoshua tells us in the recorded statement in the Gospel of Matthew, "And all things, whatsoever ye shall ask in prayer, believing, ye shall receive." As well as his statement recorded in the Gospel of Mark, "Therefore I say unto you, What things soever ye desire, when ye pray, believe that ye receive them, and ye shall have them." I will revisit here the concept of prayer and meditation and how it is synonymous. Prayer is the heart and brain working in a desire to reach the Creator. By deep meditation with the heartfelt intention to be in union with the Creator, Universe, Source, we involve our hearts and our brains, however, we must endeavor to exit the carnal mind in so doing. We must enter into the seat of Universal Consciousness, the right hemisphere of our brain while synchronizing our hearts with the Universe, resonance of feelings as we explained earlier. Feelings become so paramount, and feelings are felt in the heart. If we believe something, what ever that belief is, there is a requisite feeling associated with that belief. Therefore, Yehoshua is simply asking one to enter into deep meditation leaving our carnal minds behind, with the heartfelt desire to commune with the Creator, Source, Wholeness, Ultimate Consciousness, and in love make our request of the Creator, feeling all of the feelings in our heart of the belief that is is already accomplished herein this realm while so doing.

Yehoshua's Manifestations

It is abundantly clear when reading the King James Bible, that Yehoshua was a Master. A Master of his own self, and of the reality around him. We discussed in another chapter that he was a Yogi Master, reaching the ultimate Samadhi *(to be one with the Universe)*. When a person of this realm masters there self, and is in tune with the Universe to such a level that they are connect with the Divine Source of all things while walking *(outside of meditation)* then manifestation becomes something of a normal, effortless process.

As we discuss in the Chapter on Yehoshua's miracles, we see in some cases, it is apparent that he meditated at great length prior to a miracle. In other cases, we have no recorded information on the setting of the healing, except that he may have been inside a home, or in the street, or in the wilderness. The touching of his garment was enough for some healings to occur, which guides me to the understanding that his being was vibrating, or his overall frequency was so high, immediate resonance occurred in the person touching him because the persons desires matched that of the frequency of the Universe in Yehoshua.

The concept of passing beyond the veil, and changing the hologram projection can be witnessed through Yehoshua's life in many of his signs and wonders recorded. I am remembering the coin in the fish's mouth, or the severed ear that immediately was healed, feeding of thousand's of people with very little food, or healing of the centurion servant. All of these and many more recorded actions of Yehoshua can be seen as changes in the hologram projection that he produced.

The term crossing the veil, into the spiritual realms in order to change the hologram projection, brings to mind the recorded passage of Yehoshua's crucifixion. It is recorded in the King James Bible that the veil was torn when Yehoshua was hung on the cross. It was the Mystic Yehoshua that lived a daily life of example of how the power is truly within you, by the will of the Creator, to rise above this realm and live in the lofty spiritual realms. His living example, and true embodiment of the law was a lifting of the veil for everyone then and now. Upon his death, the ultimate display of loving your neighbor as your self, the veil was torn, or removed completely, meaning he had shown us everything we need and nothing remained. No more mystery to the spiritual evolution and true communion with the Creator, Universe, Source.

No more veil

That is to say, nothing remains secret to those desiring to change, or evolve spiritually, to climb the ladder back to Source, or Ultimate Consciousness. What does remain, is the illusions of self, the entrapment of the five senses, and the egoic self. Which through meditation, you can remove yourself from this trap of the five senses and can enter the more lofty realms of spirituality for such communion.

Thoughts are Frequencies

Every person is manifesting all the time. I would like to take a moment to illustrate this concept. We discuss in the chapter on Meditation, the subconscious and conscious brain. The subconscious is responsible for over ninety percent of our brain activity, while the conscious brain accounts for less than ten percent of our daily brain activity. The subconscious is primarily comprised of many programs placed there while we were in childhood and in theta brainwave states. This fact will be the basis of how we think and perceive the world around us and also how we behave in it. Now, thankfully, meditation is the perfect practice to reveal this programming and allow us the opportunity to change it.

The subconscious brain, which is running a majority of the day, is creating thoughts and those thoughts are creating feelings around every occurrence in our day. We have discussed elsewhere in this book, and will state again, that thoughts are frequencies, and feelings are frequencies. In the spiritual realm, the Universe only responds to feelings, which move us closer or further apart from the desired manifestation from occurring. So, putting this all together now, if our subconscious is creating thoughts we do not want consciously, which are creating feelings in opposition to our desires, we are manifesting but not what we consciously want. This sets the stage for the question, what are you thinking?

Hopefully, I have made the impression that our thoughts are pretty important. We are always thinking, consciously and subconsciously.

Therefore, we must intend to make congruent our conscious thoughts with our subconscious thoughts. Revisiting Yehoshua's statement on belief, if we feel the feelings of true belief in meditation but walk in our daily life thereafter not believing, we will not make manifest in our life our desires. I think it may be best summed up in this manner: when one is in harmony, or a state of frequency resonance with the Universe of all things *(our feelings matching its feelings)* without effort at all moments of our day *(awake and asleep)* our desires will flow to us without effort.

10

MIRACLES
EXPLAINED

I would like to walk through several of Yehoshua's miracles to see if we can process these events in a manner that allows us to begin to see we too are able to be as he was. In the Gospels of the King James Bible, we can find recorded 37 miracles performed by Yehoshua. What is interesting to note here is that each of these miracles performed by him were for specific purpose. We do not find record of anything performed by Yehoshua for amusement or for show. In fact, often times he would leave an area because his fame was spreading and he did not want that attention. Also, he would say to some to not say a word to anyone about what had happened to them.

In most general terms, the miracles recorded could be divided into three main categories: Power; Signs; and Wonder. Power would be described as acts performed that represented a connectedness or oneness with the Divine Creator, the Universe or Source, operating through him as if connected by conduit or channeling through himself into this reality. Signs would be described as synchronicities or miracles signifying in a more figurative manner the wonders of the Creator, the great Universe and Source. Lastly, wonder would be best described as an act or event that Yehoshua manifested to show those around him the awesome power and capability that one has through being connected to the Universe and daily communion with the Creator.

I will list the recorded miracles in chronological order of his life as recorded in the Gospels of the King James Bible below, however, we will exemplify the Mystic Yehoshua through only some of these recorded miracles for brevity sake. A curious reader may find extended study on this topic beneficial outside of this writing.

The Recorded Miracles of Yehoshua

- Water into wine (John 2:1-11)

- Healing of boy at Capernaum (John 4:43-54)

- Driving out evil spirit from a man in Capernaum (Mark 1:21-27, Luke 4:31-36)

- Healing Peter's Mother-in-law of a fever (Matthew 8:14-15, Mark 1:29-31, Luke 4:38-39)

- Healing of many others at evening (Matthew 8:16-17, Mark 1:32-34, Luke 4:40-41)

- Miraculous fish catch Lake Gennesaret (Luke 5:1-11)

- Cleansing a man of leprosy (Matthew 8:1-4, Mark 1:40-45, Luke 5:12-14)

- Healing centurion's paralyzed servant in Capernaum (Matthew 8:5-13, Luke 7:1-10)

- Healing paralytic who came down through roof (Matthew 9:1-8, Mark 2:1-12, Luke 5:17-26)

- Heals a mans withered hand on Sabbath (Matthew 12:9-14, Mark 3:1-6, Luke 6:6-11)

- Raises window's son from dead in Main (Luke 7:11-17)

- Calms a storm on the sea (Matthew 8:23-27, Mark 4:35-41, Luke 8:22-25)

- Cast demons into herd of pigs (Matthew 8:28-33, Mark 5:1-20, Luke 8:26-39)

- Heals a woman in crowd with issue of blood (Matthew 9:20-22, Mark 5:25-34, Luke 8:42-48)

- Raises Jairus' daughter back to life (Matthew 9:18, 23-26, Mark 5:21-24, 35-43, Luke 8:40-42, 49-56)

- Heals two blind men (Matthew 9:27-31)

- Heals a man who could not speak (Matthew 9:32-34)

- Heals an invalid at Bethesda (Luke 5:1-15)

- Feeds 5,000 plus women and children (Matthew 14:13-21, Mark

6:30-44, Luke 9:10-17, John 6:1-15)

- Walks on water (Matthew 14-22-33, Mark 6:45-52, John 6:16-21)

- Heals many sick in Gennesaret as they touch his garment (Matthew 14:34-36, Mark 6:53-56)

- Heals a gentile woman's demon-possessed daughter (Matthew 15:21-28, Mark 7:24-30)

- Heals deaf and dumb man (Mark 7:31-37)

- Feeds 4,000 plus women and children (Matthew 15:32-39, Mark 8:1-13)

- Heals blind man at Bethsaida (Mark 8:22-26)

- Heals man born blind by spitting in his eyes (Luke 9:1-12)

- Heals a boy with an unclean spirit (Matthew 17:14-20, Mark 9:14-29, Luke 9:37-43)

- Miraculous temple tax in a fish's mouth (Matthew 17:24-27)

- Heals a blind, mute demonic (Matthew 12:22-23, Luke 11:14-23)

- Heals a woman who had been crippled for 18 years (Luke 13:10-17)

- Heals man with dropsy on Sabbath (Luke 14:1-6

- Cleanses ten lepers on way to Jerusalem (Luke 17:11-19)

- Raises Lazarus from the dead in Bethany (John 11:1-45)

- Restores sign to Bartimaeus in Jericho (Matthew 20:29-34, Mark 10:46-52, Luke 18:35-43)

- Withers the fig tree on the road to Bethany (Matthew 21:18-22, Mark 11:12-14)

- Heals servant severed ear while he was being arrested (Luke 22:50-51)

- Second miraculous catch of fish at Sea of Tiberias (John 21:4-11)

What I am finding very interesting at this time of writing this chapter is just noticing the timing of his miracles. Nothing is recorded for us to read of prior to his return, or as some would refer to it, the lost years of Christ. We read of his story up to twelve years of age, and then thirty years of age is when we pick back up on the timeline of Yehoshua's life. All of the miracles recorded, and listed above, were after his return, or began when he was thirty years old. However, I am finding an amazing distinction in this, lets discuss outward miracles and inward miracles.

As mentioned earlier, all of the recorded miracles in the Gospels of the King James Bible are what I would like to refer to as outward miracles. However, one should really see the inward miracles and also what I may call indirect miracles. We can read of Yehoshua's life in the beginning of the Gospels, refer to the chapter on Yehoshua's Life. The records show us that he was a curious child, curious about knowledge and wisdom. He was very knowledgeable of the Torah by age twelve and was seeking more knowledge in the Temple at Jerusalem, however it is recorded that he was teaching the experts there, the Scribes. This to me is just one example of the inward miracles that were happening for Yehoshua.

The bible does not record this, however, other ancient documents and tradition hold that the Magi, or as the King James Bible refer to these men as "wise men", would routinely come back to visit Yehoshua and his Parents bringing gifts of gold and other items to help support them. This would be an example of an indirect miracle to Yehoshua. We can see this type of indirect miracle in the King James Bible as well, when we read of the Magi coming to visit Yehoshua, as well as the shepherds who came to worship the young Yehoshua. Mary, the mother of Yehoshua, also had many inward miracles, which would have been indirect miracles to Yehoshua, when we reflect on the various people who came to visit them prophesying, and the angels messages to her regarding the birth of Yehoshua. Joseph, Yehoshua's father as well had several inward miracles, that would be classified as indirect miracles to Yehoshua, when we reflect on the visions he had regarding preserving Mary from the time she was first pregnant, to the time they escaped King Herod's decree to kill all the young boys. This is all that we have recorded, however, it just feels right to me to mention that it is most irrevocably correct to assume that there would have been numerous inward and indirect miracles that would have been occurring all through the early years of Yehoshua, either to Joseph, Mary, or Yehoshua himself. I can exemplify this with my own life and the stories of others I am familiar with. Only today, some refer to it as luck, or fate, or happenstance. When a person becomes in tune with the harmony and flow of the Universe, one begins to see how all things work out for the good of that person. One begins to develop a faith, trust, and confidence in the Universe and then truly believes that all things work out for them.

Water into Wine

In beginning this next section of the chapter where I will describe the miracles of Yehoshua from a mystical inference, I would like to set the stage with the first recorded miracle. Yehoshua's first recorded miracle is the miracle of turning water into wine at the wedding. Now what is really interesting here, is that it was his Mother who asked him to do it. I ask the question, how did she know he was capable of doing such a miracle? We do not read of anything prior that would have given her the experience that he could do this type of a miracle. I postulate several things that I think help to exemplify the Mystic Yehoshua. We know that Mary was given dreams and visions of his birth to come, even the name. This to me shows that Mary was able to connect to the Universe and her intuitive self and received information from Source. This we are all capable of, but must become sensitive to it *(refer to other chapters)*. As she observed the behavior, knowledge and wisdom of her child, Yehoshua, over the years, she would have intuitively began to have "knowings" of her son's capabilities. We do not read of any recorded information in the Gospels of the missing years of his life, however, perhaps she knew where he was and what he was learning. It does not take much of an imagination to fill in the blanks here on this. Yehoshua returns from his travels abroad, reunites with his family, and the questions are asked of him. What did you do? What did you learn? Who did you meet? What did you see? And, so forth. I do not read of nor do I feel there was any reason he would have been secretive about his whereabouts and what knowledge he gained through out his journey to his parents.

We discuss quantum physics deeper in another chapter, however, it is important to discuss a few things here with regard to this miracle. Science is observing of late that all things are connected. This is referred to as quantum entanglement. Experiments have been conducted in this manner and have found that space nor time impacts this law of our universe. Additionally, they are observing that the atom is 99.9999% energy and 0.000001% matter. This has been the source of the phrase that "everything is energy". The question, why is this table hard, our why is this material this way, and so forth comes forward to ones mind when they begin to understand that fact, that everything is energy. The best answer is frequency, or resonance. All matter exists in its state of being due to the frequency or resonance particular to that state of being. It is the difference in resonance or frequency that allows one to perceive its existence. It has been said that it is the repulsive energy that allows us to sit on the chair, or rest our hand on the table and not fall through the surface of either.

Another item to discuss here would be the observer experiment, or the observer effect. Quantum physics has been working with this theory for some time, and simply put, reality as we observe it to be is effect by the observer. Without the observer present, the effect goes away. Some of you may desire to dive deeper into this further, and they may do so by reading the article in the February 26 issue of Nature (Vol. 391, pp. 871-874). In this article, the author discusses a highly controlled experiment demonstrating this observer effect conducted by researchers at the Weizmann Institute of Science. Why is this important, well Yehoshua turned water into wine. This is clearly a quantum physics observer effect in action. Every molecule in existence is observed into existence, and an observer who is aware of this concept, and has trust-faith-confidence in their ability to "observe" it into something different would be able to do so.

Alchemy, or the practice of transmuting, changing material is important to discuss here as well as we work through this miracle of water into wine. In the most simplistic explanation of this miracle, it is the transmutation of liquid, be it water, into a better liquid, be it wine. In fact, so much better, that it is recorded that the wine was so good, the people at the wedding commented on it, asking why they saved the best wine for last. Yehoshua would have had exposure to alchemy and would have been able to learn of this practice in Egypt, Greece, China and India *(see chapter on his journey).* I further exemplify the Mystic Yehoshua as an alchemist by his journey into the wilderness for forty days. In this record of his wilderness journey, we read of him being tempted to turn a stone into bread. If he was not an alchemist, familiar with transmutation of matter, why would he have been tempted in such manner.

I feel that the last thing that needs to be shared here for this miracle discussion, and the exemplification of the Mystic Yehoshua is the understanding he had of the holographic Universe. In another chapter, we discuss the Kabbalah, Torah and the Zohar. Therein that chapter I illustrate the Torah-Yehoshua connection as well as the Kabbalah-Yehoshua connection. The reason this connection is imperative for us to understand is that the concept of the holographic universe concept is wisdom contained within the Kabbalah. Further explained, the holographic Universe concept is that all matter exists in this realm as a projection or hologram. One can change the projection here in this realm, or reality, by consciously rising above this realm and making a change at the source of the hologram. This concept of changing the hologram is further explained in the chapters on meditation and Kabbalah, however I feel can be broken down into a simple to under stand process. Through meditation, develop a connection to Source, Universe, Ultimate Reality *(the place of origin of all matter in this realm)*, and with an intention that is of pure motive and congruent with Source, one can change the information in the energy field of the Universe, ultimately changing the hologram projection in this reality. The important key here is the trust-faith-confidence level one maintains that they were truly connected to Source, and that they truly modified the information in the energy field. What did Yehoshua say? He asked people before he healed them if they believed. He told them be it done to you as you believe. He also instructed people to ask in prayer *(meditation)* believing that it will be done and so it will be done.

Walks on Water

This miracle begins with Yehoshua sending his disciples in a ship out to sea and he instructed them to continue to the other side of the Sea of Galilee. Yehoshua disperses a crowd of people that come to see him, and then proceeds to go into the mountain to pray. The recorded scriptures do not specifically say, however one may infer from the recorded text in the Gospels that he was there praying (meditating) for several hours. Upon ceasing to meditate, Yehoshua realized that his disciples were out in the middle of the Sea of Galilee. All that is recorded is that in the fourth watch of the night, Yehoshua came to his disciples walking on the surface of the water. We do not know what transpired between the time Yehoshua first ceased meditated, and realized where his disciples were and when he come to them walking on the water. It is probable that he continued to meditate in order to perform this miracle. What is interesting with this miracle is that Peter *(one of the disciples)* asked Yehoshua to allow him to also walk on the water, which he did until he lost faith and began to sink. The next record we see is that Yehoshua saves Peter and they enter the ship. Now, there was a storm and the water was very rough at this time that Yehoshua appeared to them walking on the water so they were already anxious due to this weather occurrence, and it is recorded that the wind ceased upon Yehoshua entering into the ship.

So, on the surface here, one will see an apparent manipulation of the weather, and of matter. We discussed earlier on alchemy and what that would entail as well as to the origins of it, and I would suggest here again, we see a transmutation of the surface of the water. Additionally, we also see evidenced in this miracle the observer effect, or changes in the hologram. From a quantum physics level, the walking on water would require a resonance change or frequency change of the object touching the water so that it would then repel the object rather than allow it to pass through *(just as your hand cannot pass through the table or wall)*. From an alchemist standpoint, the molecules of the water could have been transmuted as well, or changed into something known or unknown to common man that would allow Yehoshua to be supported thereon. Prerequisite, elevation of consciousness and connection with Source, Universe, Divine, which we have already discussed herein this chapter and in other chapters that meditation is key in so doing. We see that Yehoshua was meditating when the disciples left initially, and we inferred that perhaps he continued to meditate thereafter prior to walking on the water. So therefore, I conclude that Yehoshua was in such an elevated state of consciousness from the communion with Source, Creator, Divine that he easily manipulated matter and reality.

Peter walking on water for a period of time before sinking is also very interesting to review and discuss. Allegorically, this part of the miracle shows us that we too can perform these actions and miracles with trust-faith-confidence in Source, Universe, Divine. It begins with a heartfelt intention that is congruent with Source, Universe, Divine. Our connection with source must be congruent with the level of intention or miracle desired, and we must be of elevated consciousness equivalent to the "task" desired to accomplish *(this can be exemplified by the record of Yehoshua telling his disciples that certain healing takes more meditation and fasting to accomplish)*. So it was apparent that Peter had been elevating his consciousness, and he had a trust-faith-confidence level congruent with the "task" at hand, walking on the water. However, he became distracted by the waves and the wind around him and he began to sink, or said differently, he began to drop in levels of consciousness back down to the current reality.

When meditation practice evolves to the point of touching Source, often times the meditator will become alarmed, or forget and begin to analyze the information they are receiving *(wether visual or auditory)* and they will return to the carnal mind *(left brain or the analytical brain)* and reenter the current realm of existence, thereby falling back to the current holographic projection. This is fairly common for people in early stages of meditation practice, and with time and experience will begin to learn how to control this phenomenon and remain in the right hemisphere of the brain. The right brain or the seat of Ultimate Consciousness, and remain elevated in consciousness such that they remain tapped into Source, Divine, Universe *(the place where the holographic projection may be altered to change reality in the current realm)*.

I feel this is a perfect time to introduce the Yogi Master. In India, yoga has been a practice for thousands of years and was originated therein. A yogi master is considered one who has achieved, through this ancient Indian practice of yoga, true enlightenment and self mastery, achieving a state of oneness with the Universe and transcending the limitations of the mind and ego. It is said that when a person reaches this level of mastery, they are able to access the Universal Power that flows through everything because of there realized connection with the Divine. This could be further expressed with the term *Samadhi (integration or union with the Universe)*. Yehoshua was a mystic, but he was also a Yogi Master along with many other things as we will investigate further. I was just sitting here thinking about this topic of Yehoshua the Yogi Master, and I was called to the remembrance of how he was referred to in many places in the recorded Gospels as "Master". Now, in terms of working for someone, the title of "Master" would be synonymous with the title of boss or manager, someone you would report to. These disciples chose to follow and learn from Yehoshua so that they could become like him. Therefore, he was there teacher, and in ancient practices, a Master was one who perfected the art of what they were teaching. Yehoshua was a Master in this sense, he perfected the practices he was teaching to his disciples or followers. I am concluding this discussion of this particular miracle with the desire for you to review the chapter on Yehoshua's travels abroad which would have included India.

Feeding Thousands

In flow with what we have been discussing recently in this book, I would like to further the miracle investigation on a metaphysical level with the different times Yehoshua would feed thousands with very little food. Without getting into the full details of the miracle, essentially, in two different accounts recorded in the Gospels thousands were fed with very little food and when everyone had eaten, the fragments they collected up *(the leftovers)* were more than what they started with. Now in both events, Yehoshua gave thanks for the food. In mystic terms, by giving thanks for the food, he gave it energy so that it would be nourishing and would be filling. We can see the power of our words and thoughts on food with several note worthy experiments, such as Maseru Emoto's rice experiment, and Emoto's further experiments using human energy and consciousness on water molecules. The findings were congruent with the statement that we are able to either positive or negatively impact molecules with our thought energy and words.

I would like to discuss the level of science of today and the discoveries that are unfolding which reveal the concept of *"we are the Universe, and the Universe is us"*. More and more, science is proving this statement to be true. We look to quantum physics and the investigation into the single atom containing the entire universe inside of it. Through analysis of the world around us and nature, we can begin to see this fractal pattern of reality that is repeating ever smaller to infinity yet always the same pattern. This concept of fractal patterns, which by definition is a never-ending pattern, can be further expressed when we look at our bodies, and the cells within, and what comprises the cells. This is how the quantum physicist concurs that the universe is inside of us. Now metaphysically, if this is truth, a practiced Master should be able to travel along any point of the fractal and be able to access the universal power that is flowing through everything *(because the universe is us)*.

So, returning to Yehoshua and the multitude that needed food. On a metaphysical level, he understood this concept of the power of the Universe is in us, and we have the ability to access it. His intention was out of love and compassion to feed those people, which was congruent with the intent of the Universe *(which is always and ever that of Love)*, and he gave thanks with a level of trust-faith-confidence that his intention was accomplished. Additionally, the concept of the infinite volume of abundance within the Universe, Source, Divine is illustrated with this miracle as well. When we offer our intention, in a heartfelt manner, and surrender the outcome by trusting and believing that our desire is accomplished while we feel grateful for this outcome before the event has even occurred, the Universe will provide but will do so in a level of abundance that we could not fathom *(just as the little food became baskets full of fragments)*. We will discuss this practice of manifesting further in that chapter, but essentially this was a manifestation of abundance.

Yehoshua's Healings

I want to discuss the healings that Yehoshua performed as a whole, not really singling any one out for the time being. In general, for those who may be unaware, the Gospels record many different occasions that Yehoshua performed healings for hundreds if not thousands of people. We do not know for sure how many since there are several accounts of multitudes around him and the records stating that he healed them apparently over quite a period of time (hence the need to feed them). The things he healed varied by the person, however, would range from blood illnesses, blindness, mames, possessions, diseases, and, ultimately death.

From a mystical approach, the Mystic Yehoshua would have been addressing each person from the level of their belief. In many biblical accounts, he is recorded to have asked them "do you believe", and, that their healing was because of their belief. He would also tell them to go and "sin" no more. What did the Mystic Yehoshua really mean when he would say to go and "sin" no more? Simply, to go about there life without remembrance of being sick (believing that they have been healed), and to change there life in such manner that the illness would not return. Science has been rapidly catching up with ancient mystical teachings regarding the body, and the physiological impacts upon our wellbeing when it comes to our thoughts and behaviors. We discuss the subconscious *programs* and the power of thoughts in other chapters, however, when Yehoshua instructed the persons he was healing to go and "sin" no more, he was advising them to change their thoughts and behaviors. Changing the very thoughts and behaviors through which the person ultimately caused the dis-ease in their physiology which lead to the illness or disease to present in their being.

Long-Distance Healings

Now, Yehoshua did not perform all of his healings on people who were directly in his presence. We can read of at least one account of a remote healing that took place when we read the account of the Centurion and his servant. In the Gospel of Matthew, the Centurion came to Yehoshua and asked Yehoshua to perform a healing for his servant who was still at home. The Centurion's servant was very sick, paralyzed and "dreadfully tormented". Yehoshua responded that he would go to the servants home and heal the man. However, the Centurion was humble and insisted that Yehoshua should not come to his house exclaiming that he was unworthy to have Yehoshua in his home. Rather, the Centurion insisted that Yehoshua simply speak a word and his servant would be healed. It is recorded in this Gospel that Yehoshua was amazed by the level of faith *(or belief)* that this Centurion had, and Yehoshua told the Centurion to go his way and trust that his request for healing of his servant would be as the Centurion believed it would be. In the same hour, we read that his servant was healed remotely.

What is this concept, from a metaphysical, and mystical level of remote healing? How was Yehoshua able to conduct such a healing remotely? Quantum physics has been studying the concept of quantum entanglement. Quantum entanglement is best defined as a connection between two particles even though these self same particles are separated by great distances. In ancient Indian practices, this same concept can be seen through the practice of reaching Yogi Master, or *Samadhi*. We discuss *Samadhi* elsewhere in this book, however for the purposes here, we will share again that when one reaches *Samadhi*, they are connected to all things within and without the universe. *Samadhi* allows one to see their connection with the grand scheme of the Universe and understand how all things are truly connected on an energetic level. We illustrate how Yehoshua was a Yogi Master and how he would have been knowledgeable of *Samadhi* and we see this principle exemplified in the Centurion's servant remote healing.

In a more simple example of how everything is connected, we can utilize the example of the power of our thoughts and how when someone thinks about a person with a heartfelt desire to connect with them, all of a sudden that person calls them. This is an example of quantum entanglement. The Kabbalah illustrates this principle of entanglement as well. We discuss in the chapter on the Kabbalah that we are all creatures created by the Creator, and we are returning to the Creator, Ultimate Consciousness. When we were created by the Creator, we were given a fragment of the Ultimate Consciousness. The Kabbalah tells the student that they are to correct there wrong desires, and each corrected desire returns the soul closer to oneness, or Ultimate Consciousness. Additionally, the Kabbalah teaches that as each wrong desire is corrected in one soul, every soul is corrected at the same time. Once again, this is another display of quantum entanglement, a connected universe at all levels energetically.

Anyone is capable of the practice of healing remotely, or long-distance healing for another soul. Simply put, as one develops the practice of meditation and is able to elevate themselves into the realms of higher consciousness, that person will be able to connect with the the Universe at the energetic level. It is at this level that one can impart a change in the energetic field of the other person. This is best described in the Kabbalah as the will to bestow. Now, the recipient must believe that this is possible or the energetic shift at the quantum level of the Universe may not be received by the recipient *(as stated in the Kabbalah, the will to receive)*. Through my journey's, and learning of metaphysical and esoteric knowledge, I have personally witnessed people not only healing themselves in this manner, but other people performing the healings through the remote healing process as I have described above.

I would like to take a little time here to discuss in detail, the healings that I have either participated in or witnessed firsthand in current day and time. We often read of healings, miracles, and wondrous things of old *(in antiquity)*, but the question remains, do these things happen now? Are people witnessing these things today, just as people witnessed these things over 2000 years ago with the Mystic Yehoshua. The answer to all of the above questions is a resounding, Yes.

There are many recent publications one could reference to see miracles of current date and time. I am going to highlight one publication here, "The Power of Eight, by Lynne McTaggart". In this publication, the author discusses her own remarkable findings from decades of study and research to better understand this concept of mystic healing. As she describes, utilizing group focus with their intentions on a single target, a "powerful collective dynamic emerges that can heal longstanding conditions". In her book, McTaggart discusses the miracles she has witnessed as well, in her lifetime. Now, this publication looks into the combined effects of group focus, however, as we see illustrated in Yehoshua's life, and with other ancient accounts of Spiritual Masters, more than one person performing the healing is not necessary, albeit, highly effective.

There are many different techniques in recent publications, and authors are purporting these as their systems. However, I would lovingly suggest that all things are being brought into remembrance. All modalities lead back to the one Creator, one Source, Wholeness, Oneness. We must transcend this realm, escape the entrapment of our five senses and egoic mind, elevate ourselves in consciousness, and have routine communion with the Creator, Ultimate Universe, Source. When we routinely connect with Source, we will become increasingly more whole ourselves. Each time we elevate our consciousness in intentional meditation practice, we see and perceive the world a little different, less from our five senses and increasingly more as the Creator sees the world. We will begin to develop an additional sense, which will allow us to decipher the true interconnectedness of the Universe and all of the Universes energetic pathways. This is where the mystical becomes the norm for the Mystic.

11

BENEDICTION

We have come to the final chapter of this writing, and I feel it appropriate to conclude with a benediction of sorts. Not of normal use of the term, but rather a salutation, an encouraging request between temporary departing of souls, of one Mystic to another. Go and be curious. Follow your heart. Let the known doctrines be tested with true knowledge, and seek ever loftier perspectives, and elevated states of consciousness.

We have worked our way through the various chapters here in this writing, and it is noteworthy to encourage you to reread sections herein as new knowledge and insights come to you. You will find, with new perspective, all previous writings you may have read become "new" again.

Conclusion

While not singling out any specific chapters herein this writing, I would like to provide a wrapping up of all we have discussed. It all begins with a desire within a person to know if *"there is anything else in this life"*. A desire that is born out of all the earlier desires in one's life failing to bring pleasure any longer. Out of this *new* desire, to seek more knowledge births the spiritual seed within us.

This spiritual seed, albeit, small in its beginning, will grow to ever larger size consuming the entire being of the person in time. We must allow it to grow, by so doing, we heed the importance of the seed and preserve it. The enlightenment we receive from the lowly, small seed of spiritualness, will promote more seeking and growing in divinity within the person.

The entrapment of our five senses will continue to limit our ability to seek and grow in Divinity unless we are able to cross over the *veil of secrecy*. This intentionally placed veil exists inside of every person, no person is excepted from this veil. Intentionally placed to allow the seeking person to elevate themselves above the lower realms of this life, into that of the spiritual realms. The practice of denying the energies, emotions, and thoughts of the lower animalistic centers, those psychological centers also known as chakras, one can move into the heart center. It is in the heart center that one develops the beginning of the sixth sense, or a spiritual sense for truth. Others have referred to this sense as a *knowingness* or a *resonance* within them. I have also referred to this sixth sense as a remembrance of things we have forgotten to be true. As we become more in tune with this forgotten truth, or as we are brought into ever further states of remembrance, or otherwise said, further develop this sixth sense, the veil of secrecy is lifted for us in ever increasing amounts. Not all the way at first, but in time, little by little so we can contemplate and accept the truth into our state of being. This is necessary and by the true grace of the Divine, Ultimate Consciousness, because our natural body, the egoic self is needing to be altered and in the current state will attempt to resist the truth.

The egoic self is merely a learned state of being. It is a *me first* type of attitude seeking to preserve one's state of being through utilization of the memories of the past. This egoic self, is the true definition of the fall from grace, when the Creator desired to create a creature, and imputed within this creation a desire to receive for our pleasure. When this pleasure ceases to please, we seek more desires to find pleasure from. In this seeking, we determine that spirituality is our next step in seeking pleasure. Seeking and finding pleasure in the desires of spirituality create an elevated state of energy within us that moves one out of the lower animalistic psychological centers, lower three chakras, and into the heart center. The ego cannot exist above the lower three centers, therefore there is a dying process that must carry on of the ego self. This dying process can be difficult to got through for some people, and is where the trials of the soul emerge, or the burning of the Phoenix rises from the ashes from whence it burned.

Who is the spiritual self that will rise from the ashes? Who is the new person that will be free from the egoic self? How do I continue to raise the veil of secrecy? How does one come into a state of *knowingness* and *remembrance* of truth forgotten? The answer is in meditation.

Meditation is the key to passing between worlds. These worlds being the natural realm of existence and the spiritual realms. Connecting to our subconscious is only possible when we meditate. As a person desires to evolve spiritually and become elevated in consciousness, meditation becomes a desired practice. A practice is best used to describe meditation as typically the person is not very skilled at it firstly. One will naturally fight against the ego, the carnal mind, and the body in order to become more proficient at meditation. The successful meditation is one where we commit to going through the motions, not defined by the outcome. Daily meditation will continue to put one in communion with the spiritual realms, communion with Divine, Ultimate Consciousness, and the outcome will be ever increasing Divinity and Spirituality in the meditator's life herein this realm.

Meditation can be of many varying types, but must always include a removing of oneself from this realm through escaping the five senses. Through maintaining a healthy meditation practice, one will begin to find themselves connecting to the Ultimate Consciousness without being in there normal mediation place. Divine intuition, given as needed at the appropriate times, because of our desire to evolve and because of our relationship that has been fostered with the Ultimate Consciousness by a practice in meditation.

What may feel uncommon now, or uncomfortable, as one begins a meditation practice, will soon become to feel comfortable. Contrarily, what used to feel comfortable to a person before meditating, will begin to seem uncomfortable. This shift is the evidence of the evolution of soul and spirit into ever loftier realms of perception of the Ultimate Consciousness. A practitioner of meditation will begin to realize and understand the next steps of their evolution, and be willing to give up their beliefs or comforts of the body in order to obtain. It will not seem as suffering, or they will be more willing to suffer because they are understanding the beauties that await them.

With this new understanding of suffering, the conscious observance of the sacral secretion based on lunar calendars becomes not a chore but rather a joy. Preservation of the sacral seed, made manifest through our vessel being *prepared* allows for the seed to be crucified and the gift of enlightenment becomes evident in our life. We begin to embody the truth and knowledge of *something dying* so that something *new* may arise in us. This understanding will give birth to the understanding and love for the kundalini energy within us.

The kundalini energy remains as a coiled serpent within our lower psychological energy centers, or chakras. Serpent is the term to describe, fitly so, as the serpent is anciently known as a symbol of power and ultimate wisdom. A serpent is also known in ancient times to be the feminine power within us. Just as snake charmers would use frequencies to train a serpent to rise from a basket, so too does the serpent within us rise with elevated frequencies. This term of kundalini rising is in reference to our true Divinity, true wisdom, and true power *(given to us by the Creator)* coming to life within us. It begins to rise when we begin to rise above the lower three psychological centers, the Root-Sacral-Solar Plexus Chakras. The kundalini energy is also referred to as a serpent because it rises in a coil-like fashion similarly to a serpent. The reason for this coil-like rising is the kundalini energy is passing along our spinal curvature on the vagas nerve system in an orbital manner.

This *serpent*, the kundalini energy rises and settles within us. Upon rising, it begins to *consume* as any serpent would, metaphorically the beasts within us, those animalistic desires that need to be removed so the spiritual evolution may continue. It is by the grace of the Creator, Ultimate Consciousness, that this process is a slow one, allowed to take time for the person to integrate. It is of my opinion that one should never provoke the kundalini energy to awaken, but rather allow it to occur at the right time for you. Hence the reason I have placed this topic here in the sequence of the summarization.

It should be noted that all of the afore mentioned things, or practices, are truly ongoing, in conjunction with each other, never taking the place of the other. As with the Ultimate Consciousness, Source, Universe, all things work in harmony, not dominion. You will begin with the desire to *know* more, this will never end, rather will be fostered by your ever increasing desire to meditate. Through which, the meditator will enjoy the fruits of the veil being lifted and having insight, or knowledge granted. Through a peace not provided to them of this realm, but rather of the spiritual realm through connecting by means of meditation, you will begin to preserve the Divine seeds, the sacral secretion. Through each cycle of preservation and enlightenment, you will continue to integrate the knowledge into your life through continued meditation. The kundalini energy will awaken at the appropriate time, aiding in the transmuting of the egoic self within the meditating individual. Again, this process is a continual process, cyclical in nature working to bring the practitioner into ever higher realms of spirituality. Which ultimately results in bringing our bodies into harmony.

It is in this state of harmonious presence, being in harmony with the Creator, the Universe, Ultimate Consciousness, you will find true peace, true fulfillment, true freedom. It is in this state of harmony that you truly understand the flow of energy in all things. It is in this state of harmony that you are reborn as the Spiritual Deity that we were created to be. It is in this state of harmony that you become *like* Yehoshua, the Mystic Christ.

It is with much love that I conclude this writing for the time being. I am sure, intuitively *knowing*, that I will be asked to share more in due time. For now, my heartfelt desire is that you begin to seek those ever loftier realms of spirituality, and that you endeavor to truly know yourself through meditation.